SURFACE ANATOMY

AN INSTRUCTION MANUAL

Second Edition

SURFACE ANATOMY

AN INSTRUCTION MANUAL

Second Edition

John V. Basmajian, M.D.
Professor of Medicine and Associate in Anatomy
McMaster University School of Medicine
Director of Rehabilitation
Chedoke-McMaster Hospitals
Hamilton, Ontario, Canada

428 East Preston Street
Baltimore, MD 21202, U.S.A.

Made in the United States of America

Library of Congress Cataloging in Publication Data

Basmajian, John V., 1921–
Surface anatomy.

1. Anatomy, Surgical and topographical—Handbooks, manuals, etc. I. Title. [DNLM: 1. Anatomy—Laboratory manuals. QS 25 B315s]
QM531.B27 1983 611 82-21907
ISBN 0-683-00359-3

Composed and printed at the
Waverly Press, Inc.
Mt. Royal and Guilford Aves.
Baltimore, MD 21202, U.S.A.

PREFACE TO SECOND EDITION

The amazing success of the first edition of this little book has prompted us to produce this improved new edition. Users of the earlier work will immediately note the change from its previous utilitarian format to the present improved appearance and size. It is a student's book; hence it must be convenient for students. The many new illustrations by Diane Abeloff enhance both the appearance and the presentation of information.

As before, the book is meant to be either a self-instruction manual (if necessary) or a class manual where instructors of physical examination or of living anatomy help the students. In both cases, the material covered is meant to be very useful for all students who need to know what lies at and under the body's surface.

In addition to acknowledging the contribution of Mrs. Abeloff, I must thank the staff of the Audio-Visual Department of the Chedoke Division of Chedoke-McMaster Hospitals who accepted the commission of laying out the text and figures and to Williams & Wilkins, in the person of Toni Tracy, who facilitated its completion.

McMaster University
Hamilton, Canada 1982

J.V.B.

Hamilton, Canada 1983

J.V.B.

PREFACE TO FIRST EDITION

This manual is for students and teachers of gross anatomy and physical examination. It provides a well-tried course of instruction in the only anatomy -- visual and palpable -- that most health professionals will employ after they have forgotten many details of internal structure. For almost three decades, I have taught surface anatomy as a regular part of gross anatomy, using classroom handouts for instruction. Initially, they were developed at the request of my old chief, Prof. J.C.B. Grant, when I was a young instructor in his department. Later they were modified almost annually through the fifties, sixties and seventies. Students always performed the exercises on each other with an experienced instructor to guide them. Thus, the mimeographed 'sheets' were meant to be only an aid for both students and instructors.

These instructions became so widely recognized that it became increasingly important to make them available at low cost to a much wider readership. Students and professors both complain that large topographical anatomy books with their multitude of glossy photographs are too detailed and expensive for most students. Those books also generally lack the element of student self-instruction. The present little manual is meant to be in sharp contrast to them. It can be used both in formal classes and for self-instruction.

Fortunately, one of my students, Shirley Jackson, was happy to help by making simple line drawings to illustrate many of the points raised; these illustrations supplement figures from **Primary Anatomy** and **Grant's Method of Anatomy** by permission of the publisher and the author of this present work. Thus the book was born after a gestation period of almost 30 years. The unfailing support of my secretary and chief assistant, Arlene De Bevoise, was an essential ingredient without which the book would not have been delivered. I am also grateful to the Williams & Wilkins Company, personified by Sara Finnegan, for their enthusiastic response to my enquiry as to whether they saw a need and a market for the book. Their warm support made it possible, indeed imperative, that we pursue publication of the book expeditiously.

Emory University
Atlanta, 1977

J.V.B.

CONTENTS

INTRODUCTION

Surface anatomy is an essential part of the study of gross anatomy for every student of the health sciences. For many -- probably most -- of them, the only anatomy they will ever 'see' again will be through inspection of their live patients and palpation of their deeper structures. To be able to know just what lies under the skin and what is palpable gives clinicians an immeasurably great strength in diagnosis and treatment. Firm foundations laid in the early years of anatomy (and soon after in courses of physical examination) will prevent uncomfortable floundering in subsequent years.

The only good way to learn surface anatomy is by studying living subjects -- yourself and as many companions as will submit willingly to your study. Fellow students are obvious subjects. Also refer frequently to a skeleton and to dissected material when available.

In **formal surface anatomy classes,** an experienced instructor with a group of eight to 14 students can rapidly check up on the accuracy of everyone's efforts. Generally, students should come to class prepared to disrobe partially. Thus, with work on the lower limb, both my male and female students come prepared by wearing swim trunks under their street clothes. With the upper limb, thorax and abdomen, my male students simply shed their shirts; women in mixed classes come to class wearing appropriate swim-suit halters beneath easily shed blouses. Decorum, which often is temporarily lost at the start of the first such class, is rapidly restored as the students become engrossed in the learning problems. (Students who wish to keep their shirts on -- a tiny minority -- are never discriminated against.)

Everyone participates actively and the instructor ensures that all students have succeeded with each problem before moving on. Generally, a class lasts about 45 minutes and covers about half of one of the five regions, i.e., a formal program includes not more than 10 sessions of 45 minutes each. A total of only 450 minutes devoted to **live anatomy** will illuminate the anatomy learned over many hours at the side of the cadaver. **Anatomy is the subject of living human beings and cannot be learned by studying the dead exclusively.**

With the information being learned concurrently from lectures, laboratories, programs and books, students using this manual should be able to follow the instructions and answer the questions, expecially with the aid of an instructor. As a rule, palpate with the tips of the fingers (middle and index, or index and thumb). Keep your nails short and do not prod; your subject will react with annoyance and muscles will get so tense that you will not feel underlying structures. Get your subject 'actively' relaxed. This simply means comfort and reasonable warmth. In some cases, the subject's recumbency is mandatory if you are going to successfully palpate deeper structures.

A 'skin pencil' is useful for marking out outlines of structures. Either a grease-pencil (glass marker) which is not too stiff or an eyebrow-pencil will do. **Your main tools are your eyes and fingers** -- along with a mounted skeleton (if available) and these instructions. Good learning!

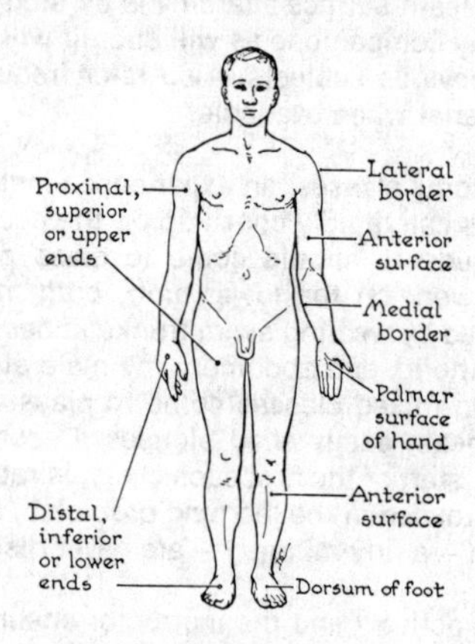

The Anatomical Position
(Except for the right forearm, which is pronated)

The **anatomical position** is the formal position in which the human body is assumed to be for all descriptive purposes. The subject is then standing erect with the eyes, palms of both hands and toes all directed forward. The palm of the hand is the anterior surface. The dorsum or back of the foot actually faces more upward and forward but is so much a part of our language that this discrepancy is overlooked.

The student should be thoroughly familiar with the terms of relationship (e.g., medial vs. lateral) and the planes of the body before starting this program of surface anatomy. It is also essential that the program be integrated with the knowledge of the parts involved through independent study.

THORAX

1

1 In the midline, palpate the **suprasternal notch** at the upper end of the manubrium sterni and between the prominent medial (sternal) ends of the clavicles. Slide your fingers down the midline for 5 cm (2") until they encounter a transverse ridge where the manubrium meets the body of the sternum at the **sternal angle** (of Louis). Slide your fingers down the midline about 10 cm (4") more to the next transverse ridge, the lower end of the body of the sternum -- the **xiphisternal junction.** Their vertebral levels are: lower border of T2; T5; disc between T9-10. Palpate the **xiphoid process** gently (it may be tender) and then the **trachea** in the suprasternal notch. (fig. 1.1)

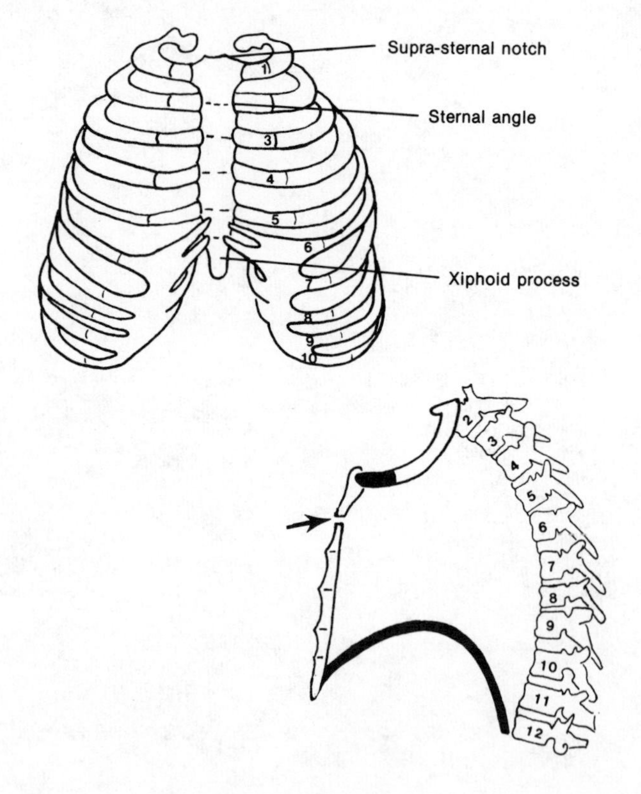

Figure 1.1 (Items 1,2)

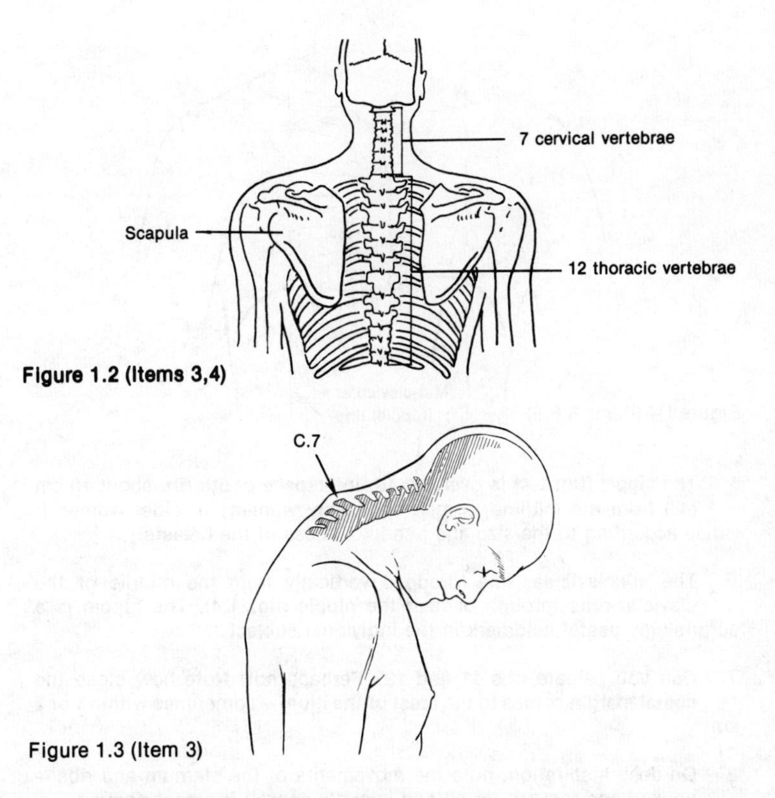

Figure 1.2 (Items 3,4)

Figure 1.3 (Item 3)

2 The **second costal cartilage** is at the level of the sternal angle. Palpate it and enumerate the ribs downwards. Which costal cartilage reaches the xiphisternal junction?

3 Turn the subject around and palpate and enumerate the **thoracic spines** (fig. 1.2) (spinous processes) starting from the 7th cervical ('vertebra prominens') which usually is the first prominent bump when palpating from above downward (fig. 1.3). Make it prominent by having the subject hang his head. Spinous processes of T1 and T4 are easy to feel (and see); those of T5 to T12 much more difficult.

4 Feel the **vertebral border** of the scapula. Note that it crosses ribs 2-7.

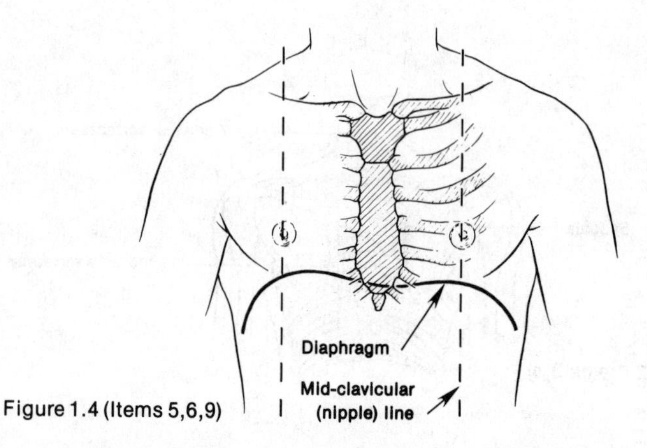

Figure 1.4 (Items 5,6,9)

5 The **nipple** (fig. 1.4) is over the 4th interspace or 5th rib, about 10 cm (4") from the midline in men and many women; in older women it varies according to the size and pendulousness of the breasts.

6 **The midclavicular line** (dropped vertically from the middle of the clavicle) runs through or near the nipple (fig. 1.4). The nipple is a surprisingly useful landmark in the individual subject.

7 Can you palpate ribs 11 and 12? Perhaps not. Note how close the **costal margin** comes to the **crest of the ilium** -- sometimes within 1 or 2 cm!

8 On deep inspiration, note the movements of the sternum and ribs -- upward and forward for all and laterally as well for most ribs.

9 The **diaphragm** (fig. 1.4). Draw its outline: the **right dome** (and liver) rises to just below the right nipple; the **left dome** (and fundus of the stomach) to 2 or 3 cm below the left nipple. In the midline, the **central tendon** is at the level of the xiphisternal junction. During vigorous inspiration the diaphragm can descend by 5 or 10 cm, forcing abdominal structures downward.

10 Draw the outline of the diaphragm on the back. (The inferior angle of the scapula is slightly lower than the nipple.) Knowing the vertebral level of the xiphisternal junction (item 1), you can plot the level of the central tendon.

PLEURA AND LUNGS

11 **Line of pleural reflexion** (where the parietal pleura is reflected from the costal cartilages or ribs onto the mediastinum or diaphragm). Draw a line from each sternoclavicular joint to the midline at the sternal angle. Carry the right line down to level of the 6th costal cartilage, then swing it laterally to the 8th rib in the midclavicular line, 10th rib in mid-lateral line and 12th rib at its middle. The left line is similar except that at the 4th c.c. it swings laterally to the left border of the sternum (mnemonic: 2,4,6,8, 10, 12) . See figure 1.5.

12 The line of reflexion on the right side passes from the back of the xiphoid to the 7th costal cartilage, i.e., an abdominal incision carried into the notch between the xiphoid and the right costal margin could pierce the right pleural sac.

13 On both sides of the line of reflexion is below the medial half of rib 12 in the **costovertebral angle,** i.e., an abdominal incision here could also enter the pleural sacs.

14 The **apex of the lung** rises as high as the neck of the 1st rib, i.e., above the clavicle, filling the **cupola** of the pleural sac.

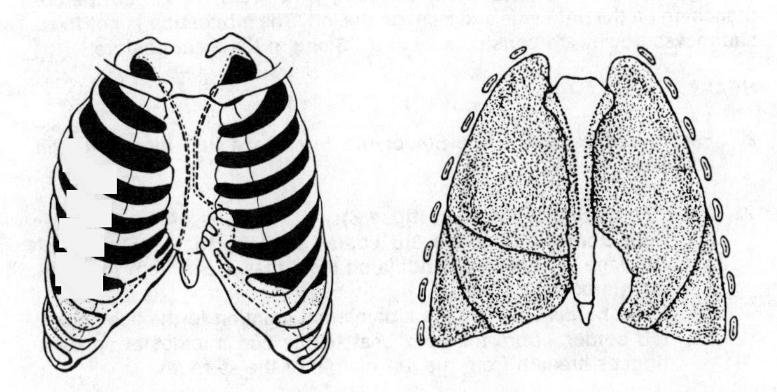

Figure 1.5 (Items 11-18)

15 The **anterior borders of the lungs** follow the lines of pleural reflexion except that a 'bite' (cardiac notch) is taken out of the left lung below the level of the 4th costal cartilage. This leaves the costo-mediastinal recess of the left pleura sac into which the sharp anterior edge of the lung can expand during vigorous inspiration; otherwise it is empty (a potential space).

16 Map the **lower borders of the lungs.** They run 2 rib-levels higher than the pleural reflexions. This leaves the right and left **costo-diaphragmatic recesses** empty gutters between the costal parietal pleura and the diaphragmatic pleura (into which the lower edge of the lung can advance with deep inspiration).

17 Main (oblique) fissure. Draw it from the 2nd thoracic spine behind to the 6th costal cartilage some 5 cm from the midline. This line coincides (by chance) with the medial (vertebral) border of the scapula when the arm is raised (hand on head) -- this is a convenient method of locating it.

18 **Horizontal fissure of right lung:** along the 4th costal cartilage from the anterior border to the oblique fissure. If you place a hand with its upper edge across the right nipple, it will cover most of the **middle lobe of the right lung.**

19 **Bifurcation of trachea:** At or below the sternal angle and slightly to the right. Map the **main bronchi** obliquely downward for 2.5 cm before branching on the right side and 5 cm on the left. The bifurcation is not fixed and moves downward considerably (e.g., 3 cm) in the erect posture.

HEART

20 Feel the **apex beat** in the 5th or 6th interspace just medial to the midclavicular line.

21 Draw the outline of the heart (fig. 1.6):
- (a) **right border** - from the 3rd costal cartilage to the 6th costal cartilage, a finger's-breadth (a bit more than 1 cm) from the right margin of the sternum.
- (b) **lower border -** across the xiphisternal junction to the apex beat.
- (c) **left border** - from the apex beat to the 2nd intercostal space a finger's breadth from the left margin of the sternum.

22 Fill in the outline of the **right atrium and ventricle**, the **left ventricle and the left auricle** of the left atrium, which is otherwise a posterior structure forming most of the 'back' of the heart.

23 Draw the great vessels:

(a) the **right brachiocephalic [innominate] vein, superior vena cava** and **inferior vena cava,** above and below the heart along the right margin of the sternum.

(b) the **left brachiocephalic vein** running obliquely from the left sternoclavicular joint to join the right brachiocephalic vein behind the 1st costal cartilage.

(c) the **aorta** arching up to the middle of the manubrium sterni.

(d) the **pulmonary trunk [artery]** and its right and left branches 'within' and below the aortic arch.

24 **Cardiac valves** could be marked on the chest wall along an oblique line from the left costal cartilage to the midline just above the xiphisternal junction in the following order: pulmonary, aortic, mitral, tricuspid, but --

25 The 'heart sounds' of the above valves (picked up by the stethoscope) are best heard on the chest wall at sites other than the above (this will be explained in clinical medicine.)

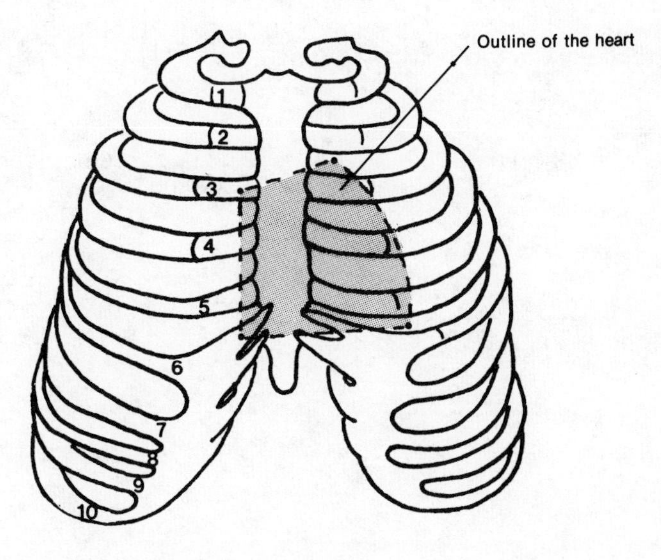

Figure 1.6 (Items 20-25)

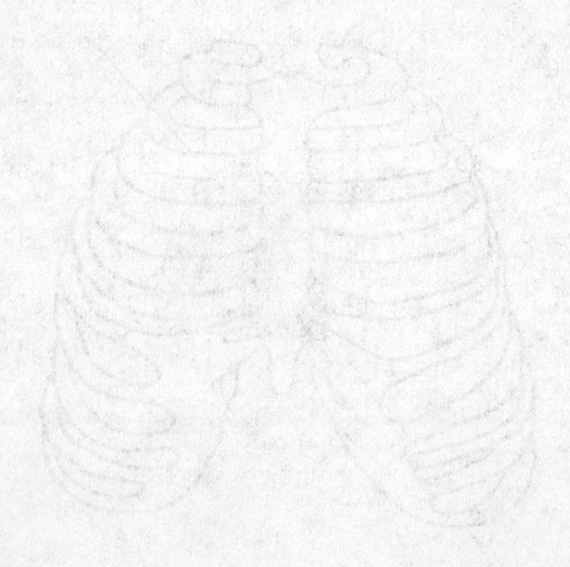

ABDOMEN

2

Palpate, indicate, or map out:

1 **Iliac crest:** it is subcutaneous and palpable throughout its length (fig. 2.1). **Anterior superior iliac spine:** to palpate it on yourself, place your hand on the iliac crest and, by dipping and hooking the fingers, press the fingertips laterally, upwards and backwards. To feel a subject's A.S. spine, use the tips of the index and middle fingers.
Iliac tubercle: the most lateral and, therefore, the highest point seen from the front. To palpate it, press upwards with two fingers.
Highest point of the crest: Join the highest points of the right and left iliac crests with a towel edge across subject's back. This will cross the 4th lumbar spine, an important level in spinal anesthesia (in which the physician wishes to avoid the lower end of the spinal cord at the 2nd lumbar level (L.2).
Posterior superior iliac spine: in the visible dimple at the level of the 2nd sacral vertebral spine. This is the lowest limit of the cerebrospinal fluid in the subarachnoid space.

2 **Pubic tubercle and pubic crest:** you can feel these on yourself or a patient quite easily, even through light clothing.

3 **Inguinal ligament:** usually neither palpable nor visible, this underlies the groove between the abdominal wall and the thigh. In some men it is palpable at its medial end.

4 **Costal margin:** lower 6 costal cartilages (review). Can you feel the free floating ends of the 11th and 12th costal cartilages? How close are yours to the iliac crest?—probably only 2 or 3 cm.

5 **Lower end of the body of the sternum** (xiphisternal joint) is felt as a transverse ridge on pressing upwards between the ends of the 7th costal cartilages.

6 The **linea alba,** often a visible furrow between the right and left rectus abdominis muscles, is broad above the navel and narrow below it, i.e., the medial margins of the rectus abdominis muscles diverge above.
Lateral border of the rectus: a curved line from the pubic tubercle (fig. 2.2), crossing the costal margin a hand's-breadth from the median plane (near the tip of the 9th costal cartilage) and on up to the 5th costochondral junction. **Lying supine,** note the movements which cause the recti to contract. Try raising the head and shoulders and/or both the lower limbs. When you raise only one lower limb, why do you find a difference?

7 **Junction** of the fleshy and aponeurotic parts of the **external oblique** is a line that curves from near the anterior superior iliac spine upwards and medially to cross the costal margin at the lateral border of the rectus abdominis.

8 **Direction of fibers of internal oblique:** what is the direction at the level of A.S. Spine? -- below the A.S. Spine? (see fig. 2.3.)

9 **Deep inguinal ring:** a finger's breadth above the middle of the inguinal ligament (generally not palpable).

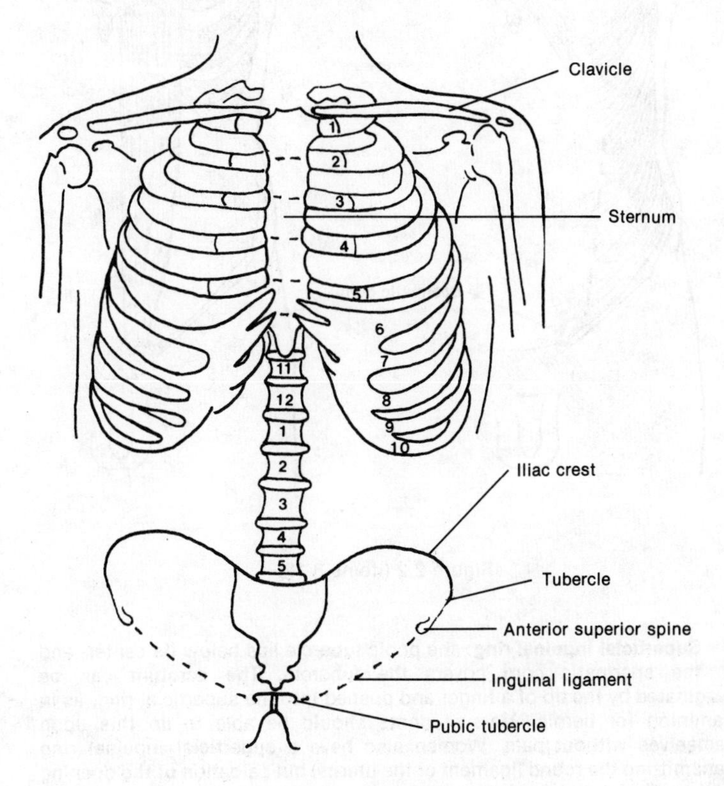

Figure 2.1 (Items 1-5)

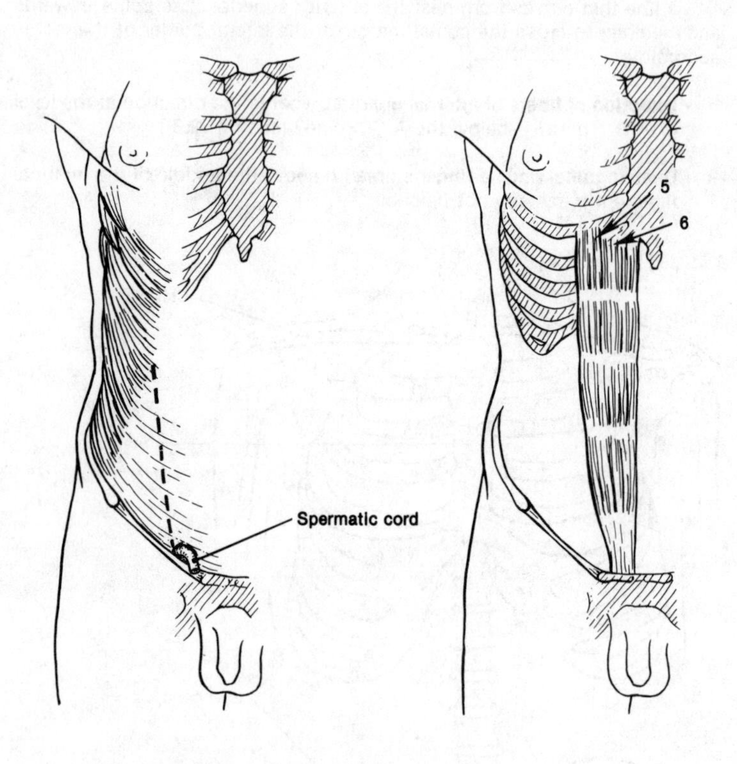

Figure 2.2 (Items 8-12)

10 **Superficial inguinal ring:** the pubic tubercle lies below its center, and the spermatic cord covers the tubercle. The scrotum can be invaginated by the tip of a finger and pushed into the superficial ring, as in examining for hernia. Male students should be able to do this upon themselves without pain. Women also have a superficial inguinal ring (transmitting the round ligament of the uterus) but palpation of the opening is generally futile.

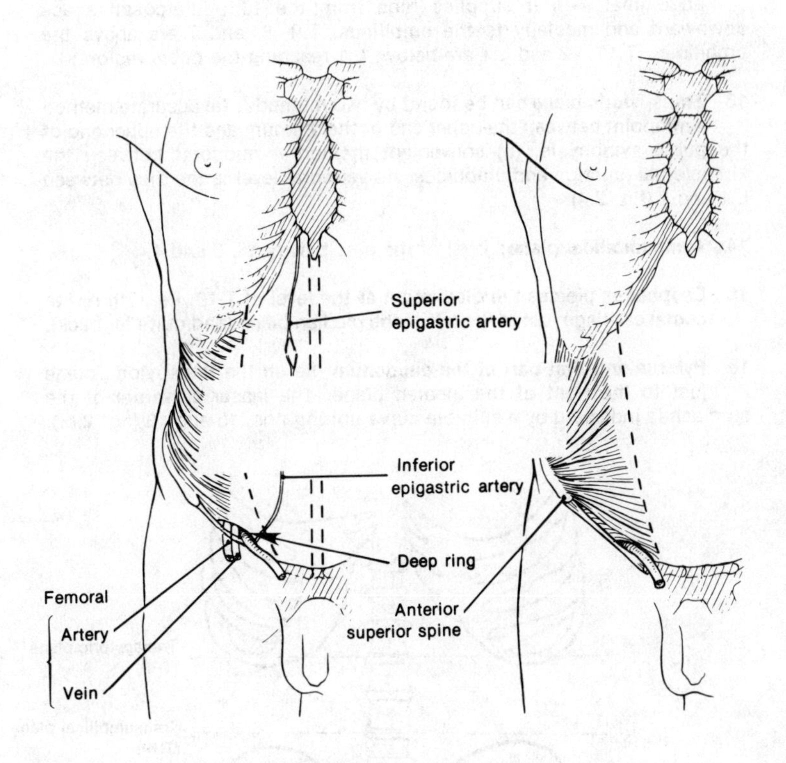

Figure 2.3 (Items 6,8)

11 **Inferior epigastric artery:** it runs from the external iliac artery towards the umbilicus, but running vertically behind the lateral third of the rectus abdominis.

Medial to the artery lies the **inguinal [Hesselbach's] triangle,** the site of a **direct inguinal hernia.** Lateral to it is the deep inguinal ring, the site or origin of an indirect hernia (which recapitulates the course of the descent of the testis).

12 The **10th thoracic nerve** courses to the umbilicus. The strip of abdominal skin it supplies runs from the 10th intercostal space downward and medially to the umbilicus. T.9, 8, and 7 are above the umbilicus; T.11, 12 and L.1 are below, L.1 reaching the pubic region.

13 **Transpyloric plane** can be found by two methods: (a) accurate method --midpoint between the upper end of the sternum and the upper end of the pubic symphysis. (b) convenient method -- midpoint between the xiphisternal junction and umbilicus. Its vertebral level is the disc between L.1 and 2 (fig. 2.4).

14 **Transumbilical plane:** level of the disc between L.3 and 4.

15 **Esophagus** pierces the diaphragm at the level of T.10, i.e., 7th or 8th costal cartilage just to the left of the median plane (and quite far back).

16 **Pylorus and first part of the duodenum** lie on the transpyloric plane just to the right of the median plane. The **lesser curvature** of the stomach is indicated by a suitable curve uniting nos. 15 and 16 (fig. 2.5.).

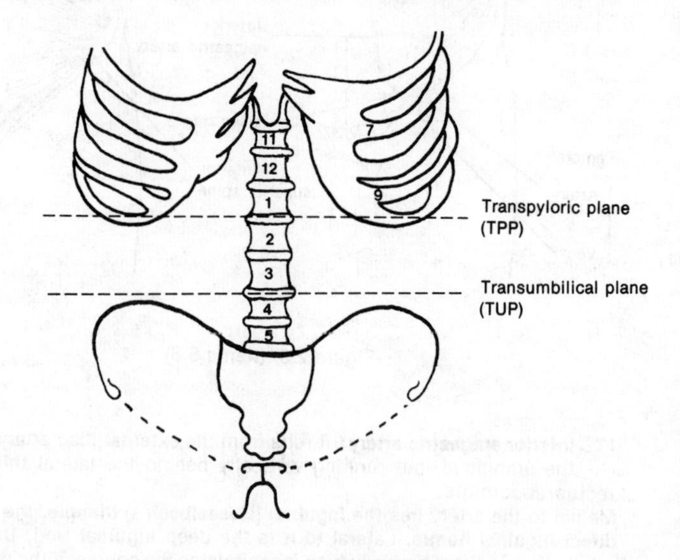

Figure 2.4 (Items 13,14)

17 **Duodenojejunal junction** is just below the transpyloric plane just to the left of the median plane. The two ends of the duodenum are, therefore, not very far apart.

18 Where the duodenum crosses the aorta it overlaps the origin of the **inferior mesenteric artery.** This artery arises as far above the umbilicus (2 cm) as the aortic bifurcation is below the umbilicus.

19 **First coil of the jejunum** passes downwards and to the left in front of the lower end of the left kidney. The **last coil of the ileum** passes upwards and to the right across the pelvic brim to the ileocolic orifice.

20 **McBurney's point,** ⅓ of the distance from the right anterior superior iliac spine to the navel, is of clinical importance (appendectomy). It overlies the **vermiform appendix.** The **ileocolic orifice** lies just above.

21 **Root of the mesentery** runs from the duodenojejunal junction to the ileocolic orifice. (It crosses the transverse (3rd) part of the duodenum, (aorta), inferior vena cava, right ureter, and psoas muscle.)

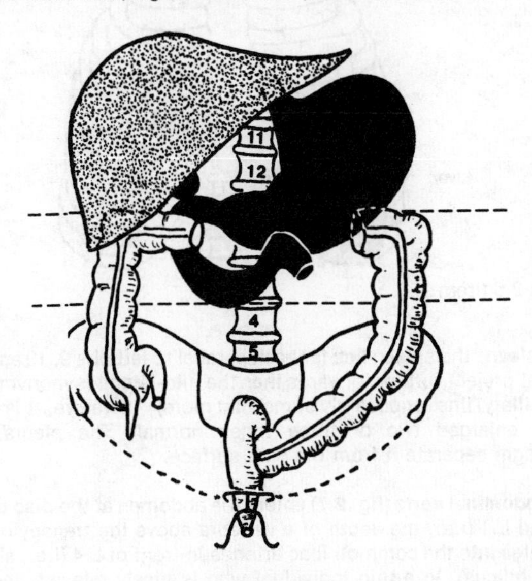

Figure 2.5 (Items 16-22)

22 The fundus of the **gall bladder** lies at the lateral border of the rectus abdominis below the costal margin and normally is not palpable. Below the neck and body of the gall bladder lie the superior part of the **duodenum** and the **transverse colon.** Although the latter is called 'transverse', it is often 'u'-shaped and dependent, and often approaching or entering the pelvis.

23 **Liver:** in the right mid-lateral line (half-way between the anterior and posterior midlines of the torso) it lies opposite ribs 7, 8, 9, 10 and 11. It curves to just below the right nipple, and, crossing the midline, continues to reach a point somewhat farther below the left nipple (fig. 2.6). It is depressed by the heart in the median plane to the level of the xiphisternal junction. Its sharp lower border crosses the pylorus and the fundus of the gall bladder in the transpyloric plane; this margin of the normal liver is palpable only occasionally.

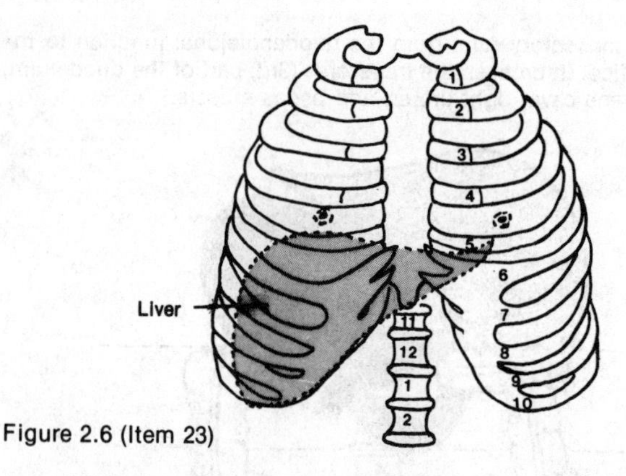

Figure 2.6 (Item 23)

24 **Spleen**: the spleen lies far back parallel to left ribs 9, 10 and 11; it does not project further forwards than the mid-lateral (synonymous with the mid-axillary) line (or possibly somewhat more); therefore, it is not palpable unless enlarged two or three times normal. The pleura, lung, and diaphragm separate it from the skin surface.

25 **Abdominal aorta** (fig. 2.7) enters the abdomen at the disc between T.12 and L.1 (i.e., the depth of a vertebra above the transpyloric plane). It bifurcates into the common iliac arteries in front of L.4 (i.e., slightly below the umbilicus). In a thin individual who is nicely relaxed, the pulsations may be felt quite easily.

26 **Common and external iliac arteries:** a curved line from the aortic bifurcation to the midinguinal point (i.e., midpoint of a line joining the anterior superior iliac spine to the symphysis) traces the course of this main arterial trunk heading for the lower limb. This places it just medial to the deep inguinal ring (see item 9).

27 **The internal iliac artery** descends into the pelvis minor (true pelvis) from the main trunk at the junction of its upper ⅓ (common iliac) and lower ⅔ (external iliac).

28 **External and common iliac veins** lie within the bifurcation of their arteries (i.e., medial to them).

29 **Inferior vena cava,** 2 or 3 cm wide, lies to the right of the aorta and is wider; it arises in front of the body of L.5.

30 **Celiac trunk [artery]** arises from the aorta as it enters the abdomen (disc between T.12 and L.1). The **crura** of the diaphragm sit astride it (fig. 2.8).

31 The **celiac plexus** surrounds it. The **celiac ganglion** lies on its side. From the ganglion, nerves pass laterally to the (**medulla** of the) **suprarenal gland** (fig. 2.8).

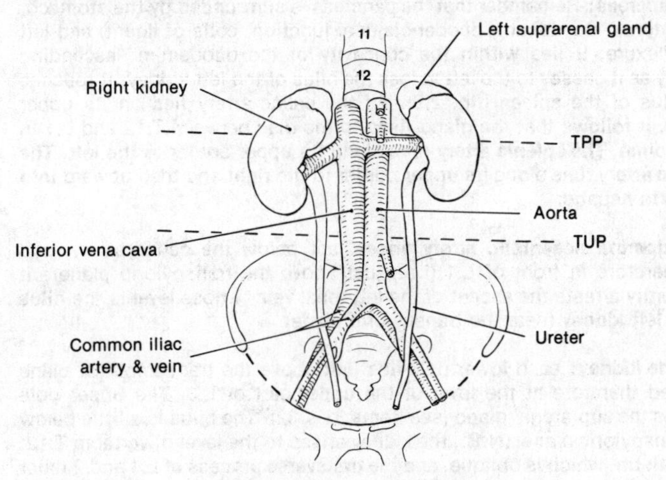

Figure 2.7 (Items 25-29, 34-35)

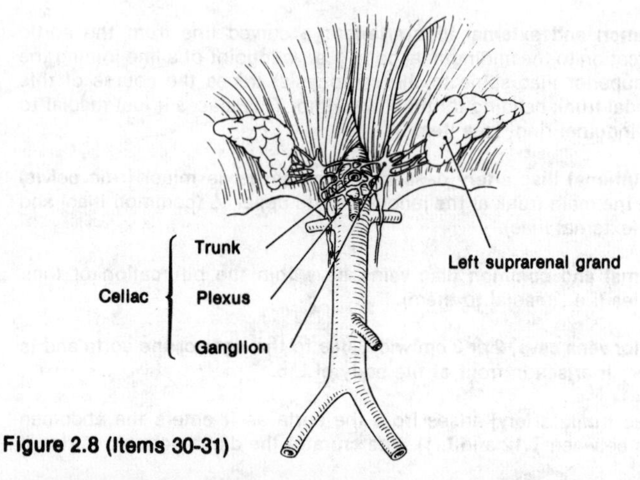

Figure 2.8 (Items 30-31)

32 **The suprarenal gland [adrenal],** therefore, lies in front of T.12 and L.1. It is not far from the median plane. The **upper pole of the kidney** must, therefore, reach the level of T.12.

33 **Pancreas:** Remember that the pancreas is surrounded by the stomach, pylorus, duodenum, duodenojejunal junction, coils of ileum, and left colic flexure. It lies within the concavity of the duodenum; ascending slightly as it passes to the left across the hilus of the left kidney, it touches the hilus of the spleen (fig. 2.9). As the celiac artery lies on its upper border, it follows that the gland rises to the disc between T.12 and L.1 in the midline. **The splenic artery** runs along its upper border to the left. **The hepatic artery** runs along its upper border to the right and then upward into the porta hepatis.

34 **Superior mesenteric artery** arises just below the celiac artery, and therefore in front of L.1 (i.e., just above the transpyloric plane); it apparently arrests the ascent of the left renal vein, whose level is the hilus of the left kidney (near the transpyloric plane).

35 **The kidney:** each lower pole lies just above the transumbilical plane and therefore at the level of the upper part of L.3. The upper pole reaches the suprarenal gland (see items 31 & 32). The hilus is a little below the transpyloric plane. (N.B., the kidney arises to the level of vertebra T.12. The 12th rib, which is oblique, and the transverse process of L.1 and 2 must be behind it). The kidneys are usually impalpable. (see No. 42.)

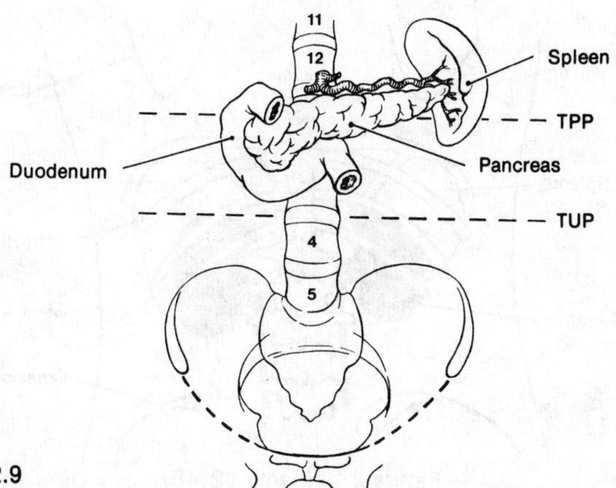

Figure 2.9

36 **Ureter:** Drop a line from the hilus of the kidney (about 3 cm from the median plane) vertically downwards to cross the external iliac artery in front of the internal iliac artery (see Nos. 26 & 27). (fig. 2.7).

37 **Testicular [or Ovarian] Artery:** arises from the aorta somewhere below the renal artery which lies behind the renal vein. (See No. 34.) The testicular artery descends to the deep inguinal ring and runs through the inguinal canal to the testis. The ovarian artery crosses the external iliac artery to reach the ovary lying on the side wall of the pelvic cavity.

38 **The testis:** If you are a male, try to palpate the epididymis -- with care, you can. How is the ductus (vas) deferens identified by palpation? -- it feels like whipcord.

39 Where would you palpate for the **lymph nodes** that drain the testis (or ovary) -- on the sides of the aorta in the abdominal cavity!

40 In clinical work, the **prostate** is palpated by rectal digital examination. The normal **female internal genitalia** are impalpable by abdominal palpation alone, usually requiring bimanual examination with the fingers of the one hand in the vagina and the other palpating through the abdominal wall above the pubis.

41 **Urinary bladder:** When the bladder is empty, the parietal peritoneum passes from the anterior wall of the abdomen onto the back of the

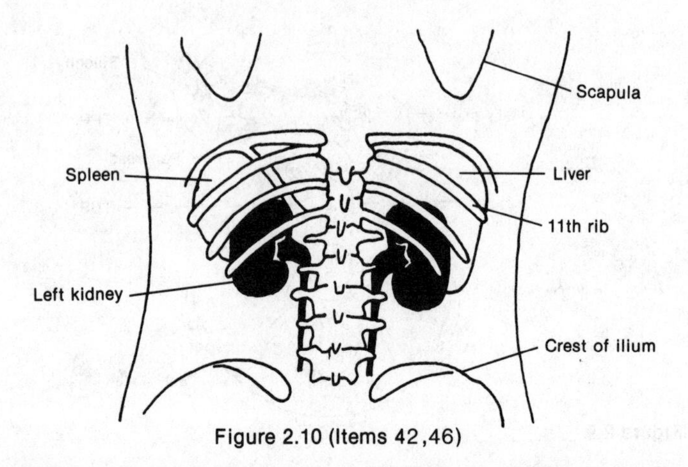

Figure 2.10 (Items 42,46)

symphysis. When the bladder is filled, the loose peritoneum gets 'stripped off' the abdominal wall some distance above the symphysis. A very full bladder can be palpated as a rounded mass above the pubis.

FROM THE BACK

42 **Lower border of the 12th rib.** How many finger's breadths above the iliac crest is the tip of your 12th rib? One, two, or more? This varies.

43 Since the usual approach to the kidney is from behind, translate the data given in item 35 to the back, thus: Its lower end is about a finger's breadth above the iliac crest and the upper end is above the 12th rib (and so there is pleura behind the upper pole of the kidney).

44 **Ureter:** From the hilus on the transpyloric plane (i.e., the disc between L.1 and 2) it drops downwards anterior to tips of transverse processes, some 3 or 4 cm from the midline (fig. 2.10).

45 **Liver:** The landmarks for its upper levels (deep in the torso) are the inferior angle of the right scapula and just below the inferior angle of the left scapula; it is depressed in the midline as it crosses the 8th thoracic spine. Its lower border follows the 11th right rib.

46 **The Diaphragm:** The upper boundary of the diaphragm and liver are one and the same. See 'Abdomen', item 23, and 'Thorax, item 11.

UPPER LIMB 3

1 Run you index and middle fingers along the **clavicle**, upper surface of the **acromion,** and crest of the **spine of the scapula.** Between what 4 muscles are you palpating? Looking in a mirror, bring your platysma (a skin muscle) into action by grimacing strongly (it is part of the facial muscles of expression) as though you are trying to loosen a tight collar. Note that the skin of your thorax is lifted.

2 Palpate the **acromioclavicular** joint by pressing medially above the acromion.

3 With the thumb and index finger palpate the **angle of the acromion** (fig. 3.1). It is an important landmark from which measurements down the arm are made by clinicians, e.g., item 22, below.

4 While carrying a weight in the hand, grasp the sloping **anterior border of the trapezius muscle** (which descends to the lateral ⅓ of the clavicle). Note that it is now prominently in action (fig. 3.2).

5 The borders of the **deltoid muscle** are usually visible when the arm is abducted against resistance. Draw the oval outline of the **subacromial [subdeltoid] bursa** which passes deep to the acromion about 3 cm and downward about the same distance deep to the deltoid. When it is abnormally swollen with fluid, a needle slipped below the acromion enters it easily. Otherwise it is collapsed.

6 On extending the elbow, the triceps muscle becomes prominent just below the posterior border of the deltoid, and its lateral head may be distinguishable from the long head.

7 While grasping the anterior fold of the axilla (fig. 3.3), determine which parts of the **pectoralis major** act during flexion (elevation), extension against resistance (against the back of a chair), adduction, and medial rotation of the shoulder. (Clavicular head: flexion; sternocostal head: extension and adduction; both heads: medial rotation of the humerus.)

8 Identify the various **ribs** by counting down from the 2nd rib at the **sternal angle** (as you did in the Thorax chapter, item 2).

9 Feel the tip of the **coracoid process** below the lateral part of the clavicle. It is in the **deltopectoral triangle** (a visible hollow in some persons) sheltered by the anterior border of the deltoid; therefore press firmly upward and laterally, deep to the deltoid.

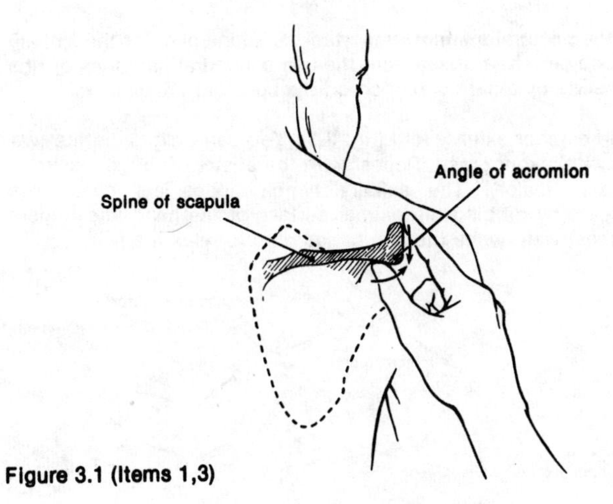

Figure 3.1 (Items 1,3)

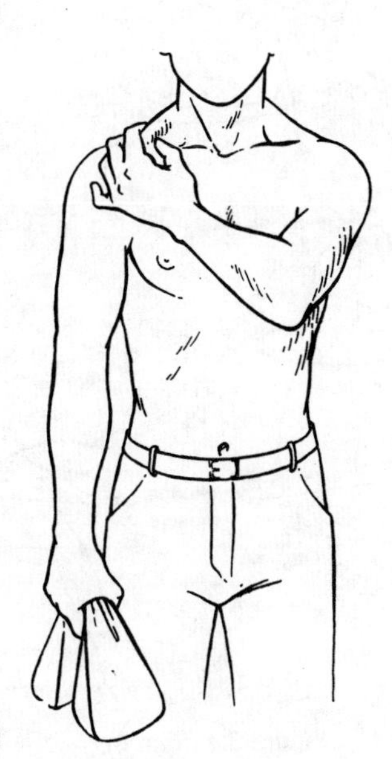

Figure 3.2 (Item 4)

10 Outline the **pectoralis minor,** important as a landmark for the axillary vessels and nerves. It arises from the costochondral junctions of ribs 3, 4, 5 and inserts by a narrow tendon on the coracoid process.

11 Grasp the anterior axillary fold (fig. 3.3). You can now name the two muscles within your grasp. Repeat with the posterior fold (latissimus dorsi and teres major). The axillary lymph nodes are sometimes palpable.(Palpate by rubbing the palmar surface of the hand and fingers against the chest wall, while the arm hangs limp to relax the fascia).

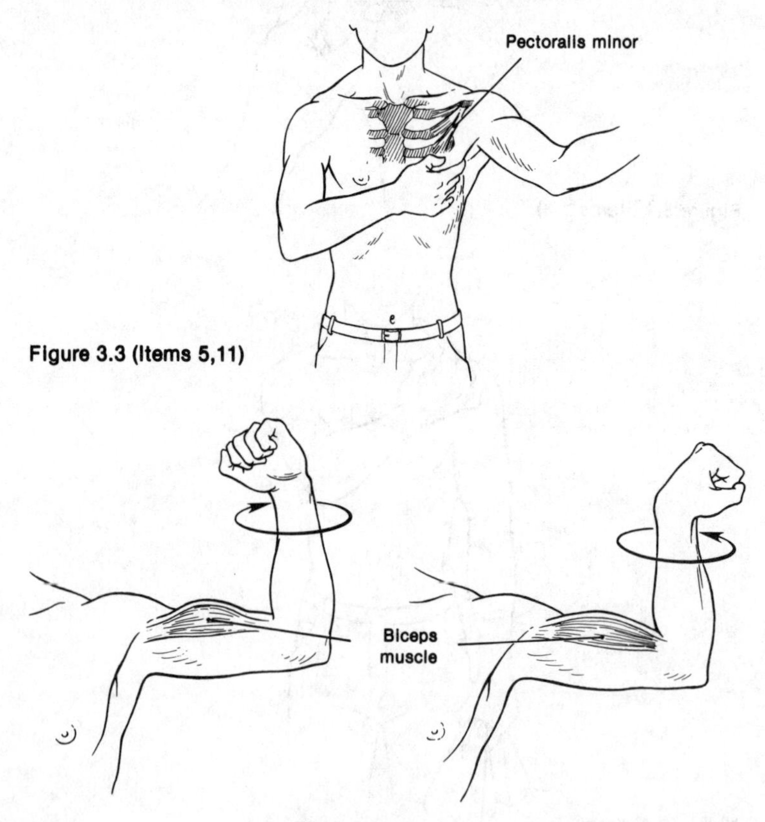

Figure 3.3 (Items 5,11)

Figure 3.4 (Item 6)

12 **Biceps brachii.** With the forearm pronated, flex the elbow and observe that the biceps swells and that on supination the swelling increases (fig. 3.4).

13 Map the course of the **axillary and brachial arteries.** Run a line from the middle of the clavicle to the middle of the cubital fossa (the hollow of the elbow). This line should pass a finger's breadth medial to the tip of the coracoid process. Note the changing relationships of the brachial artery to the humerus. Feel its pulse throughout its course.

14 The three cords of the **brachial plexus** lie around the artery behind the the tendon of the pectoralis minor, i.e., a finger's breadth medial to the tip of the coracoid process.

15 **Tubercles of the humerus.** The **greater** is lateral and projects beyond the acromion causing the roundness of the shoulder; grasp it through its covering of deltoid muscle. The **lesser** is inferolateral to the coracoid process and points forward; palpate it sliding beneath your fingers while you medially and laterally rotate the humerus.

16 The **bicipital groove** lies between the tubercles and faces forwards when the arm is in the anatomical position. Perhaps you can feel it when you feel the lesser tubercle.

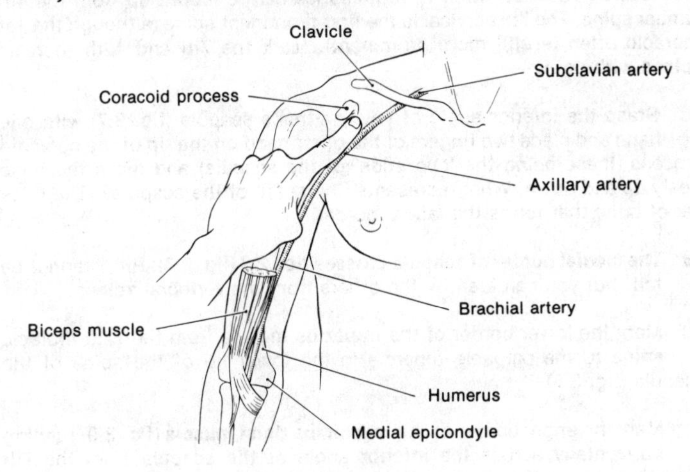

Figure 3.5 (Items 14, 15, 16)

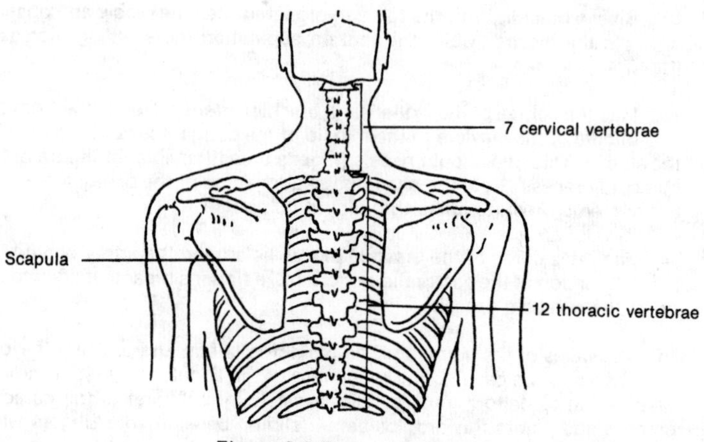

Figure 3.6 (Items 17,19)

17 Count the **vertebral spines** (fig. 3.6) down from the 7th cervical (the vertebra prominens which you identified with the Thorax, item 3). In the Abdomen Section, item 1, you also learned to count up from the 4th lumbar spine. The 7th cervical is the first prominent spine (although the 1st thoracic often is still more prominent). Mark the 7th and 12th thoracic spines with dots.

18 Grasp the inferior angle of your partner's scapula (fig. 3.7) with one hand and place two fingers of the other hand on the tip of the coracoid process (these being the 'two ends' of the scapula) and move the bone freely up and down. What represents 'the shaft' of the scapula? (The thick bar of bone that forms the lateral border).

19 The **medial border of scapula** crosses ribs 2-7 (fig. 3.6). Rib 1 cannot be felt, but you can identify the others from the vertebral spines.

20 Mark the lower border of the trapezius muscle from the 12th thoracic spine to the palpable tubercle in the lower lip of the spine of the scapula (fig. 3.8).

21 Mark the upper border of the **latissimus dorsi muscle** (fig. 3.9) running horizontally across the inferior angle of the scapula from the 7th thoracic spine to the posterior fold of the axilla (in which it was grasped earlier) (item 11).

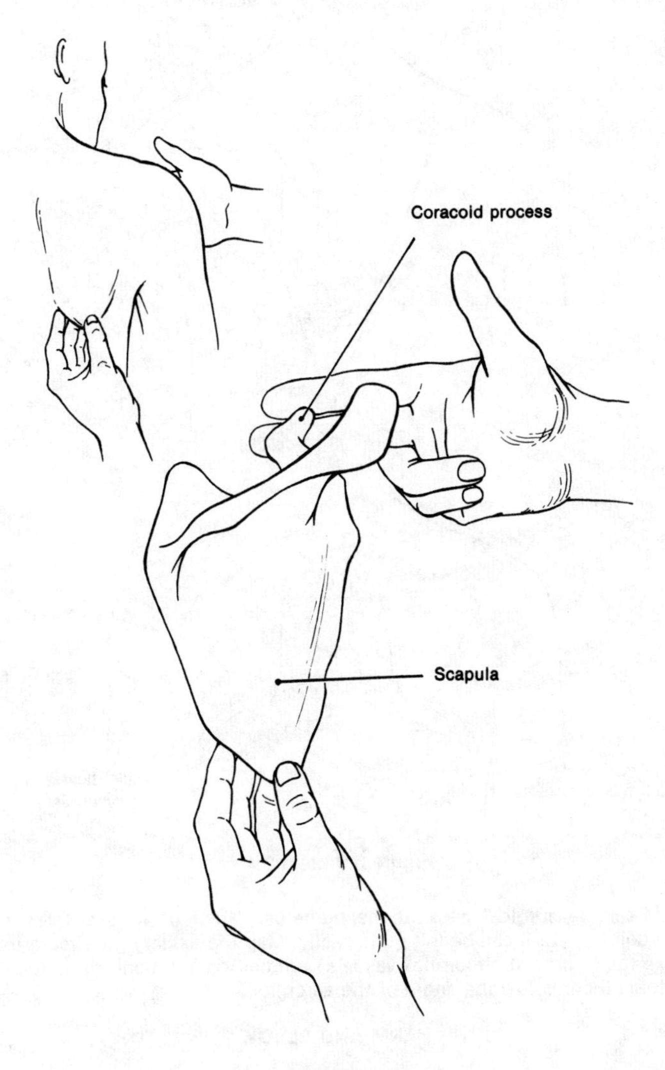

Figure 3.7 (Item 18)

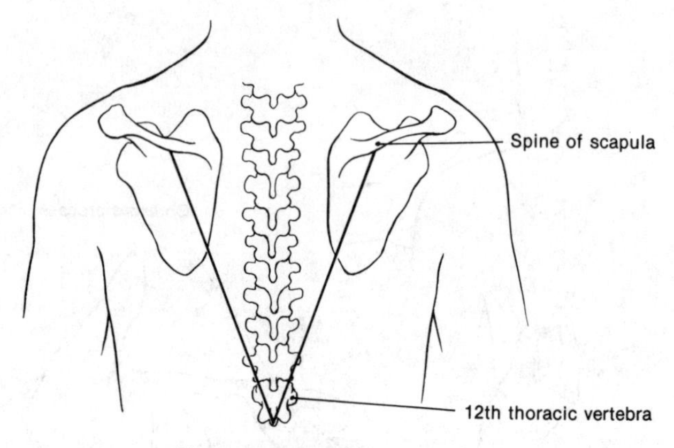

Figure 3.8 (Item 20)

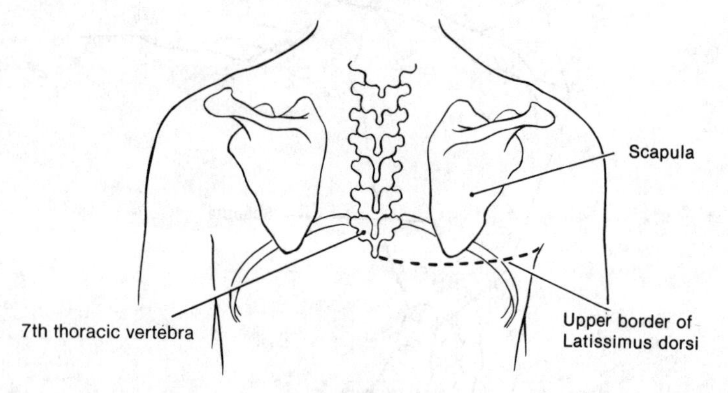

Figure 3.9 (Item 21)

22 Feel the **surgical neck** of the humerus. Much of it is covered by deltoid, but it can be felt quite easily. Map the **axillary [or circumflex] nerve** (and humeral circumflex vessels) running on the bone deep to the deltoid, 5 cm below the angle of the acromion.

ARM AND ELBOW

23 **Brachial artery.** See item 12, and figure 3.11.

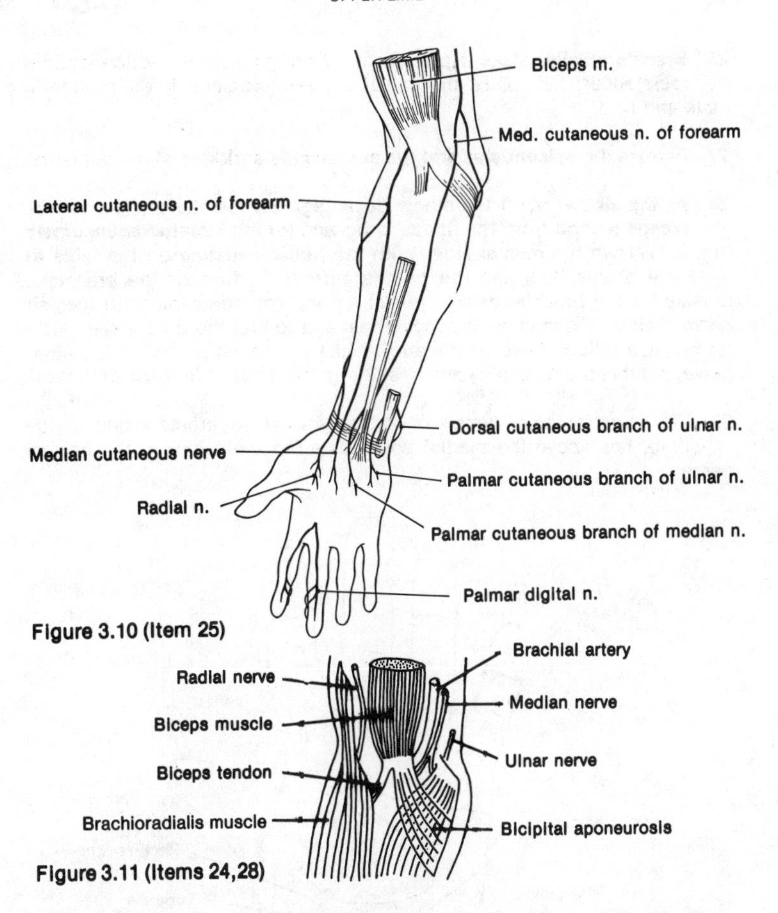

Figure 3.10 (Item 25)

Figure 3.11 (Items 24,28)

24 With your finger tips try to roll the **radial nerve** where it runs vertically in its spiral groove which runs almost vertically on the back of the humerus behind and below the insertion of the deltoid.

25 Indicate where the **medial and lateral cutaneous nerves of the forearm** become cutaneous. The line along which they emerge is constant but the height is not. Map their courses anterior to the elbow region before dividing (fig. 3. 10).

26 **Brachioradialis.** Make it prominent by bringing it into action against resistance ; i.e., place the fist of the semipronated forearm under a table and lift.

27 Palpate the **epicondyles and the supracondylar ridges** of the humerus.

28 **At the elbow:** With the elbow flexed against resistance, feel (a) the **biceps tendon** from the lateral side; and (b) the **bicipital aponeurosis** (fig. 3.11) from the medial side. With the forearm resting on the table to relax the biceps, feel and roll on the anterior surface of the **brachialis** muscle (c) the **brachial artery** pulsating and the companion (d) **median nerve.** Behind the medial epicondyle feel and roll (e) the **ulnar nerve.** How far can you follow these up the arm? When you rolled or flicked the ulnar nerve, did this produce 'pins and needles' (paraesthesia) in the little finger?

29 The **supratrochlear lymph node,** the lowest superficial node in the limb, lies above the medial epicondyle (not palpable) in the normal person.

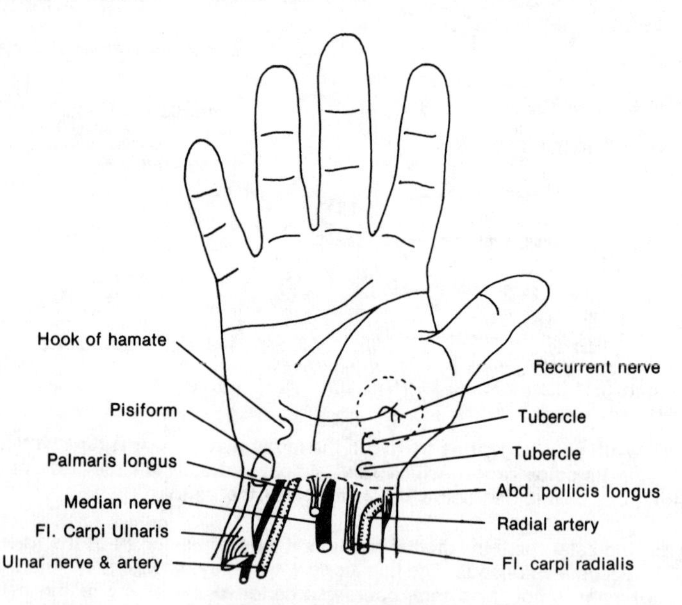

Figure 3.12 (Item 30,32)

30 In the forearm, map the radial and ulnar arteries, radial and ulnar nerves, and on the opposite forearm, pronator teres, flexor carpi radialis, palmaris longus and flexor carpi ulnaris muscles. Can you feel the pulse of an anomalous superficial ulnar artery? -- it occurs in less than 5% of limbs.

Forearm and wrist

31 Outline the **pronator quadratus muscle** (impalpable) which covers the lowest ¼ of the shafts of the radius and ulna.

32 At the **wrist** (fig. 3.12) --

(a) See the lowest **skin crease.** It corresponds to the upper border of the **flexor retinaculum** which is the size of a postage stamp with its long axis transverse.

(b) Bring into action by vigorously flexing the wrist, the **palmaris longus,** if present. Mark the median nerve deep, or perhaps lateral, to the palmaris longus. It passes deep to the middle of the flexor retinaculum.

(c) Mark the lunate bone, behind palmaris longus tendon at the skin crease.

(d) Bring into action **flexor carpi radialis** and follow it vertically across the **tubercle of the scaphoid** and down the groove of the trapezium.

(e) Feel the pulse of the radial artery, lateral to the flexor carpi radialis tendon. How far proximally can you feel your own?

(f) Bring into action **flexor carpi ulnaris** and follow it easily to the pisiform bone. Beyond that it becomes the pisometacarpal and pisohamate ligaments, which are impalpable.

(g) Palpate the **pisiform bone.** Feel the deep resistance of the **hook of the hamate.** The **ulnar artery and nerve** descend between these two bones.

(h) Bring into action **flexor digitorum superficialis [sublimis]** by making a fist and feel its tendons above the **carpal tunnel** (formed by the flexor retinaculum).

(i) **Abductor pollicis longus** tendon forms the profile of the wrist laterally.

(j) Between fingers and thumb grasp the **lower end of the radius** in order to note the crest-like anterior border of its inferior articular surface. It lies 2 cm above the lowest skin crease. Your wrist-watch strap encircles the lower ends of the radius and ulna, not the carpus (wrist) itself.

(k) Can you feel the pulse of the ulnar artery at the wrist? It is more difficult than that of the radial but not impossible.

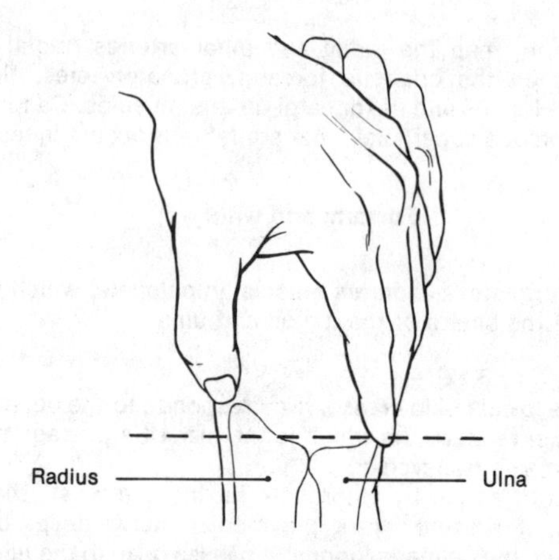

Figure 3.13 (Item 33)

33 With the thumb and index palpate the tips of the radial and ulnar **styloid processes,** noting their levels (fig. 3.13). The radial styloid is always lower by about 1 cm and this is important knowledge for the surgeon treating a Colles' fracture of the lower end of the radius.

34 **Trace the cephalic and basilic veins** from the dorsal venous arch on the back of the hand to the front of the forearm and on up the arm on the lateral and medial sides of biceps. The cephalic continues subcutaneously to the deltopectoral triangle. They may be made prominent by hanging the arm, then obstructing the venous flow by light circular pressure above the elbow.

35 In your palm note the movements that accentuate the **skin** creases and compare them to the metacarpal bones (fig. 3.14).

36 Map the **superficial palmar arch** at the level of the web of the outstretched thumb and having 1 of 3 possible endings. Do you possess a superficial palmar branch of the radial artery (1 of the 3 possibilities) pulsating where it descends across the proximal part of the thenar eminence?

37 Map the (impalpable) **deep palmar arch** on the dorsum of the hand at a level just distal to the palpable bases of metacarpals 2, 3, and 4; translate the marking to the palmar skin.

38 While moving the thumb into abduction, flexion and adduction, feel in the **thenar eminence** the changing actions of the **abductor pollicis brevis, opponens pollicis** and **flexor pollicis brevis,** and in the web of the thumb the **adductor pollicis.** The **hypothenar muscles** are less distinct although the **abductor digiti minimi brevis** can be discriminated by palpation during abduction.

39 The vulnerable **nerve to the three thenar muscles** can be covered with a dime where it recurs distal to the flexor retinaculum, 3 cm below the tubercle of the scaphoid.

40 On a finger, map the **tendon insertions** of flexor digitorum profundus to the base of the distal phalanx and flexor digitorum superficialis (sublimis) to the sides of the middle phalanx.

41 On another finger, mark the attachments of a **fibrous digital sheath.** The upper ends rise to the level of the metacarpal heads in the palm.

42 On still another finger, mark the **digital vessels and nerves** running along the sides of the fibrous sheaths.

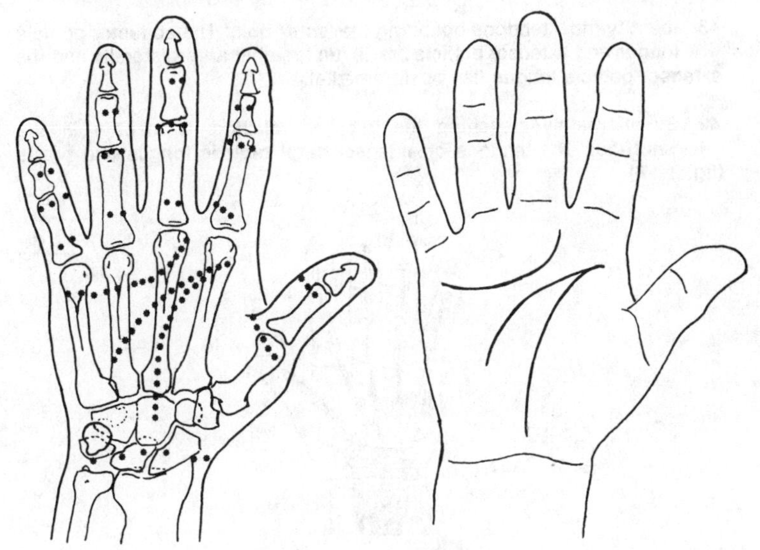

Figure 3.14 (Item 33)

43 Map the common palmar and digital sheaths (fig. 3.15).

44 On the other hand, map the usual cutaneous distribution of the **ulnar, median** and **radial** nerves (fig. 3.16). The ring finger shares all three.

45 Feel the joints of a finger in flexion. Mark them on the back of the finger. Compare the marks with the skin creases in front (fig. 3.14).

BACK OF FOREARM

46 On the **back of the forearm** palpate on the ulna: (a) the **olecranon** (a subcutaneous bursa overlies it), (b) sharp subcutaneous **posterior border,** (c) **head** and **styloid process,** (d) subcutaneous lower third of the **medial surface** (between flexor and extensor carpi ulnaris).

47 Palpate on the **radius:** (a) the **head** in the depression below the lateral epicondyle and (by pronation and supination) roll it under two fingers; (b) the soft coating of the radial shaft by the **supinator** muscle—what traverses it?; (c) the **styloid process** and **dorsal radial tubercle** (the latter is a pulley for the extensor pollicis longus).

48 Identify the 3 tendons bounding the **'snuff box.'** The abductor pollicis longus and extensor pollicis brevis run together anterolaterally and the extensor pollicis longus lies posteromedial.

49 By alternately clenching the fist and relaxing it, identify in the **'snuff-box'** the tendons of **extensor carpi radialis longus** and **brevis** (fig. 3.17)

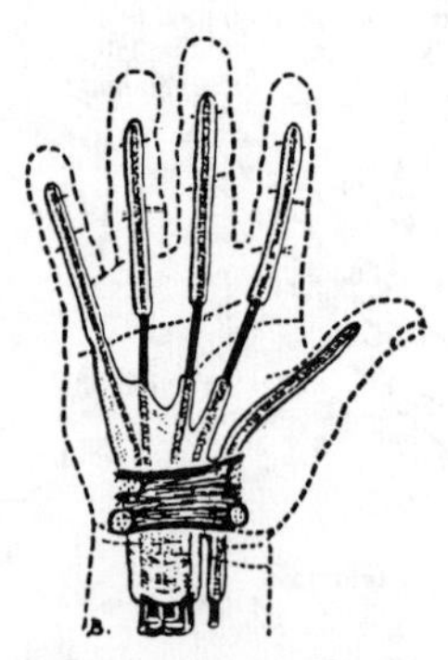

Figure 3.15 (Item 43)

50 Observe the **dorsal venous arch** crossing the 'snuff-box' and palpate the pulse of the **radial artery** to where it runs in a deeper plane to plunge through the **1st dorsal interosseus muscle.** Palpate the surface of (but you cannot identify) the **scaphoid** and **trapezium** bones in the floor of the 'snuff-box.'

51 **Extensor carpi ulnaris** fills the groove between the head and the styloid process of the ulna. It cannot be identified there, but when you adduct the hand you can feel it proximal to the **5th metacarpal base.**

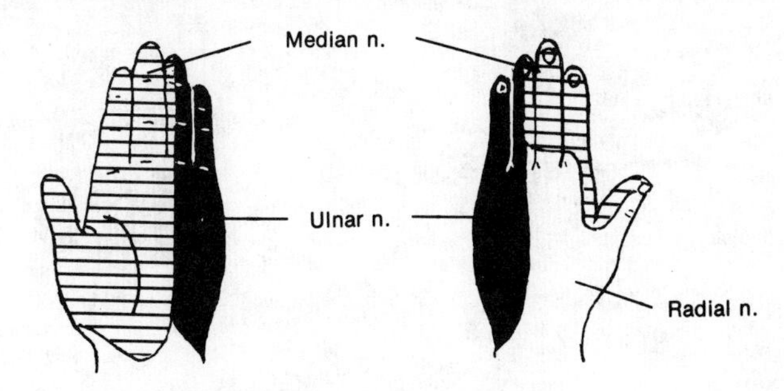

Figure 3.16 (Item 44)

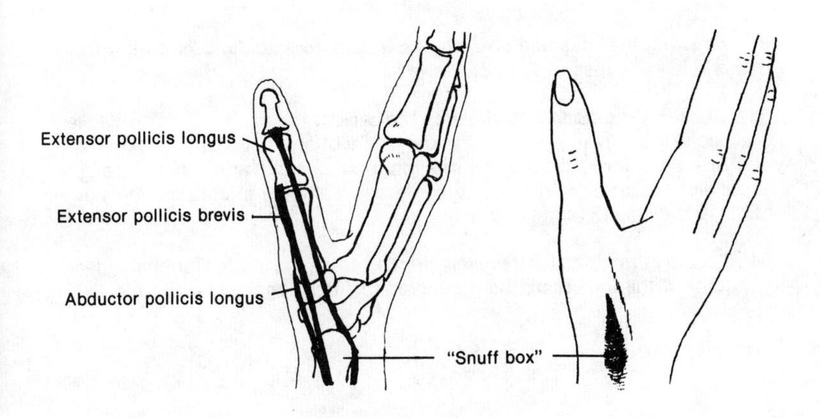

Figure 3.17 (Items 47,48)

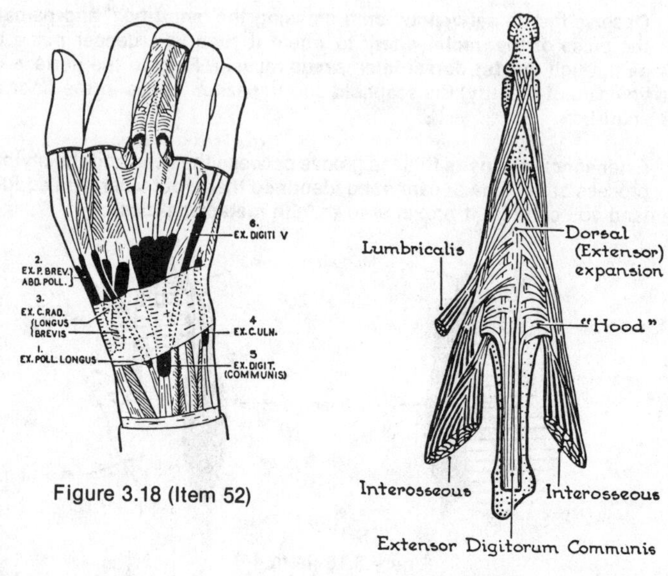

Figure 3.18 (Item 52)

Figure 3.19 (Item 53)

52 Knowing the uses and breadth of **extensor retinaculum,** plot its upper and lower edges (fig. 3.18).

53 Observe the **extensor tendons** on the back of the hand (fig. 3.19). Move each finger in turn and reveal the connections between the **extensor digitorum** tendons. Where do the tendons expand and where are the expansions inserted? Where do the interossei and lumbricals join the sides of the expansions? (see fig. 3.19).

54 Feel the **first dorsal interosseus** harden when the metacarpophalangeal joint of the index is either abducted or flexed against resistance.

LOWER LIMB

4

Palpate, mostly using two fingers, **indicate,** or **map** the following:

1 Anterior superior iliac spine (palpable) } Review

2 Pubic tubercle (palpable) } Review

3 Inguinal ligament (usually impalpable) } Review

4 **Saphenous Opening** in the fascia lata. Indicate the center of this oval about 3 fingers' breadths below the pubic tubercle. (fig. 4.1)

5 **Great Saphenous Vein.** This large vein is usually visible in thin males when standing. Trace it from the dorsal venous arch on the foot: anterior to the medial malleolus; obliquely across the lower third of the tibia; just behind the medial border of the tibia; a hand's breadth behind the medial border of the patella; to the lower edge of the saphenous opening.

6 **Inguinal lymph nodes. Superficial nodes** commonly may be palpable below and parallel to the inguinal ligament and long saphenous vein. They drain superficial parts of the courses of the saphenous and superficial inguinal and superficial epigastric veins; i.e., the limb, the skin of abdomen below the umbilicus, the penis (not the glans), the scrotum (not the testis or ovary), and the anus. **Deep nodes** lie along the femoral vein and pass through the femoral canal.

7 **Head of the Femur**. A horizontal line from the pubic tubercle bisects the head and touches the top of the greater trochanter. The psoas tendon crosses the head; anterior to it is the femoral artery -- midway between the symphysis pubis and the anterior superior iliac spine (i.e., the midinguinal point). (fig. 4.2)

8 **Femoral artery:** With the hip joint extended, you may feel its pulse at the midinguinal point.

Femoral nerve: lateral to the artery and, therefore, anterior to the iliacus.

Femoral vein: medial to the artery and, therefore, anterior to the pectineus.

9 **Adductor tubercle** is often used as a landmark. To palpate it, grasp the medial side of the lower thigh and press laterally and downwards with the radial border of the hand. On pressing the knees together, the **cord-like tendon of the adductor magnus muscle** becomes taut and palpable.

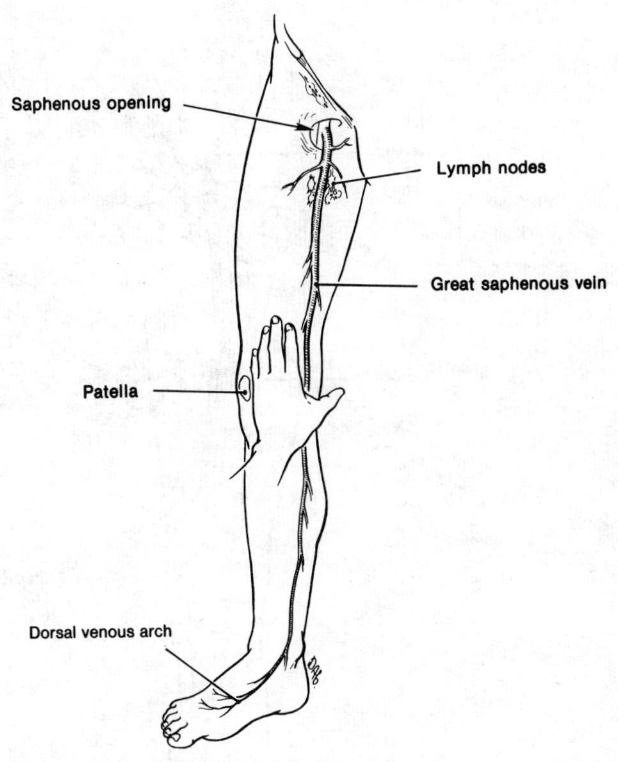

Figure 4.1 (Items 4-6)

10 **Femoral artery:** it follows the upper 3/4 of a line from the midinguinal point to the adductor tubercle; a small branch (saphenous a.) follows the lowest ¼ (the limb being rotated laterally).

11 **Profunda artery** arises 3 or 4 cm below the inguinal ligament, and lies behind the femoral artery at the apex of the femoral triangle 10 cm below the inguinal ligament.

12 **Adductor Longus** tendon arises from the pubis just medial to the tubercle. So, with the thigh abducted, palpate it and follow it toward the tubercle.

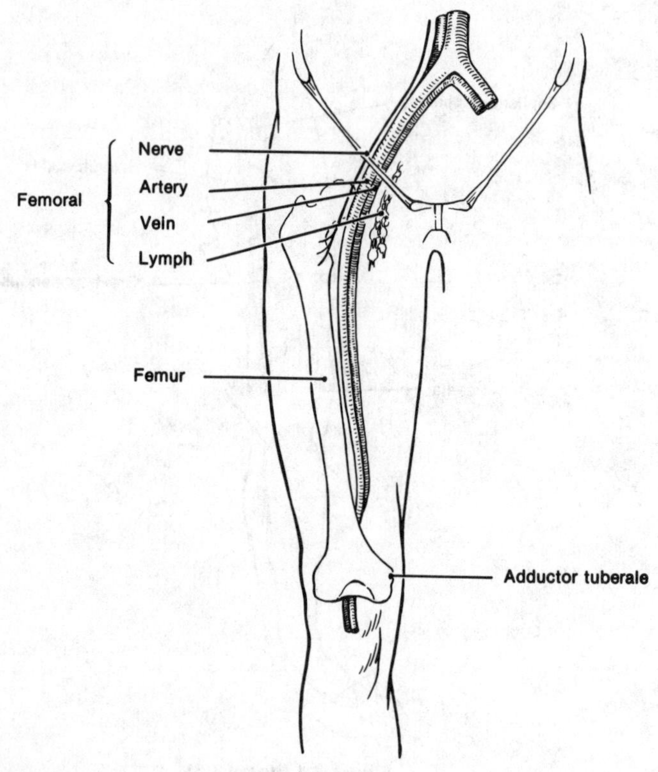

Figure 4.2 (Items 7-10)

13 **Sartorius muscle** runs from the anterior superior iliac spine to the medial surface of the tibia postero-inferior to the tuberosity. When the muscle is in action it stands out boldly where it bounds the femoral triangle, and will not be mistaken for rectus femoris which lies lateral to it.

14 Sit down, fully extend the knee, and raise the heel from the floor. Observe the hollow on each side of the patella and the quadriceps tendon. The lateral hollow is limited behind by the prominent anterior border of the iliotibial tract; its posterior border is not to be mistaken for biceps femoris tendon which lies some distance behind it.

GLUTEAL REGION

15 Palpate the **iliac crest** (fig. 4.3). The **highest part of crest** is level with the 4th lumbar spine. The **posterior superior spine** is in a dimple at the level of (a) the 2nd sacral spine; (b) the middle of the sacroiliac joint, and (c) the bottom of the subarachnoid space.

16 The **posterior surface of the sacrum and coccyx** leads down to the **tip of the coccyx** which is palpable in the midline, 5 cm above and behind the anus.

17 **Ischial tuberosity** is covered by gluteus maximus during extension of the hip but is uncovered during flexion so that it bears the weight in sitting (fig. 4.4). Palpate its lower surface and medial border during hip flexion (e.g., while sitting on your finger-tips).

18 **Greater trochanter.** To grasp it, you must **passively** abduct the limbs widely to relax the covering abductor muscles [**gluteus medius** and **minimus**]. You cannot grasp it when standing on one foot because those muscles are strongly active.

19 **Gluteus maximus** (fig. 4.5). Its lower border is a line from the tip of the coccyx, across the ischial tuberosity to the junction of the upper ⅓ with the lower ⅔ of the femur. Its upper border is parallel to its lower border -- from the posterior superior iliac spine to above the greater trochanter.

Its nerve enters its center. The favorite site for intramuscular injections is the 'upper-lateral quadrant' of the buttock to avoid the sciatic nerve. Such an injection enters either the gluteus medius or the upper part of the gluteus maximus.

20 **Piriformis.** To map its **lower border,** bisect a line from the posterior superior iliac spine to the tip of the coccyx. Join the midpoint to the top of the greater trochanter.

21 **Superior gluteal artery and nerve** appear in the apex of the greater sciatic notch above the piriformis.

22 The **other vessels and nerves** enter the gluteal region **below** the piriformis muscle. You should be able to name them.

23 **Sciatic nerve**, the most important structure appearing below the piriformis, passes midway between the ischial tuberosity and the

greater trochanter (where the gluteus maximus covers it), thence vertically down the thigh to the apex of the popliteal fossa. (The limb should be in the anatomical position). (It is most readily exposed in the angle between the biceps femoris and the gluteus maximus). It drops vertically down the middle of the back of the thigh deep to the hamstring muscles.

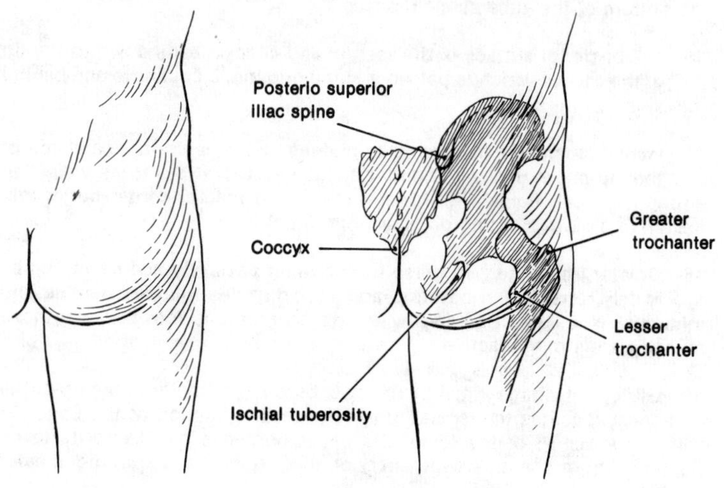

Figure 4.3 (Items 15-18)

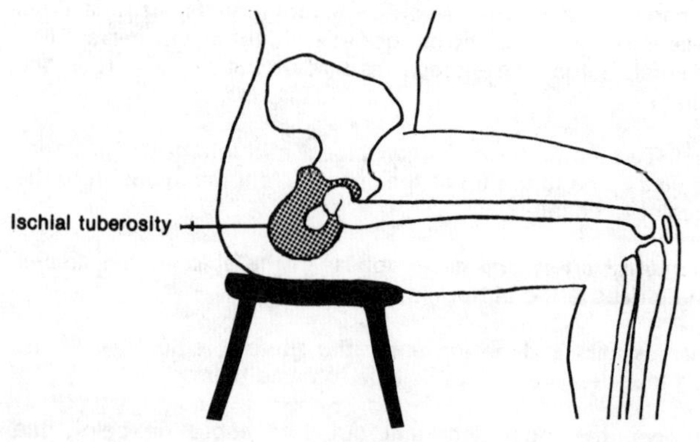

Figure 4.4 (Item 17)

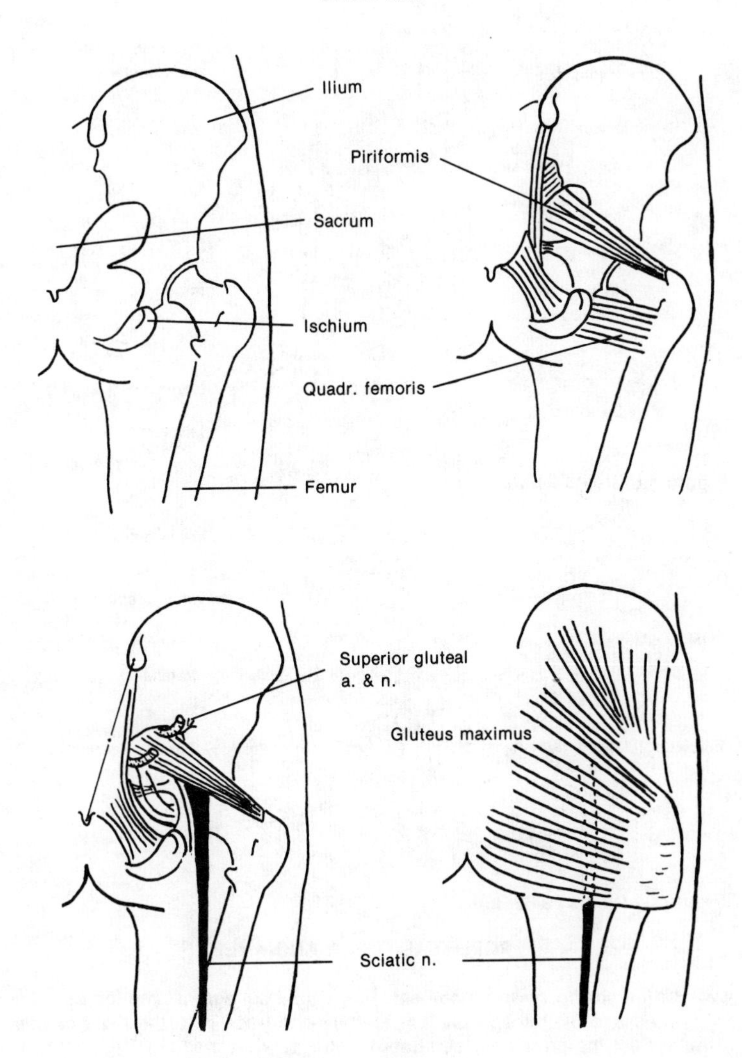

Figure 4.5 (Item 19-23)

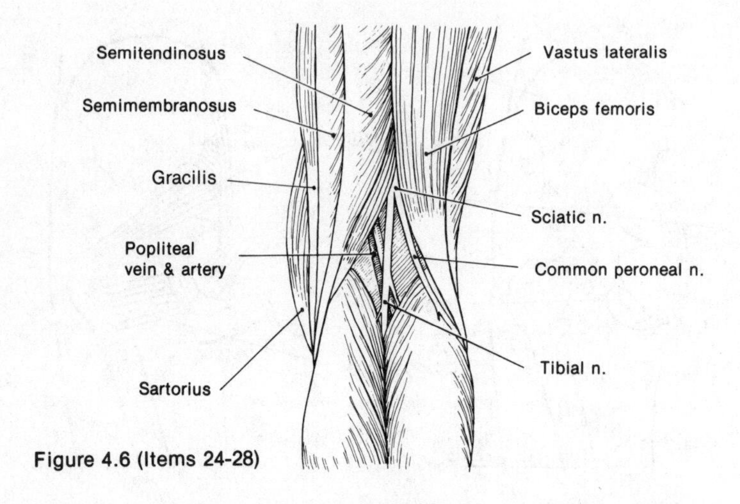

Figure 4.6 (Items 24-28)

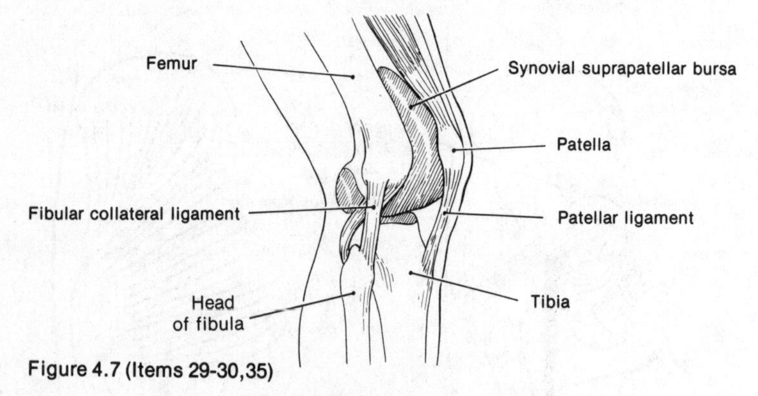

Figure 4.7 (Items 29-30,35)

POPLITEAL FOSSA AND KNEE

24 While sitting down, knee bent, press the heel against the leg of your chair and feel **biceps tendon** laterally and trace it to the head of the fibula. Feel the prominent **semitendinosus** tendon medially (fig. 4.6). It can be made to spring away from **semimembranosus** tendon because it is attached lower down the tibia (fig. 4.10).

25 **Tibial [medial popliteal] nerve** bisects the fossa vertically.

Common peroneal [lateral popliteal nerve] follows the biceps femoris which shelters it. It descends behind the **head of the fibula** where it can be felt, curves forwards lateral to the **neck** of fibula, where it should be felt, rolled, or 'flicked' with the finger-tips.

26 **Popliteal vessels** enter the fossa from the medial side and then practically bisect it vertically. The **tibial nerve** is, therefore, lateral to the vessels above and, of course, superficial to them. (Why 'of course'?)

27 **Small Saphenous Vein.** Trace it from the dorsal venous arch of the foot, behind the lateral malleolus, lateral to the tendo calcaneus and then upward between the two heads of the gastrocnemius to plunge deep in the center of the popliteal fossa.

28 **Popliteal lymph nodes** (not palpable normally) drain the course of the small saphenous vein. They lie deep to the popliteal fascia.

29 **Patella.** Palpate its margins. When standing with the body bent forward (or when sitting with the heel resting on a chair), the patella may be moved from side to side to such an extent that much of its posterior surface can be actually palpated along the medial and lateral edges. Note that it lies entirely above the level of the tibiofemoral articulation, i.e., in front of the lower end of the femur.

30 Palpate the **patellar ligament.** Note its length and breadth and follow it down to the tibial tubercle which is very prominent.

31 Palpate the margins of the femoral condyles. Note that as the knee becomes more and more flexed, first the 'trochlear' area becomes uncovered, and then the tibial surface of the medial condyle, and lastly the tibial surfaces of both condyles.

32 Palpate the **epicondyles.** They are located at the posterior center of an ellipse. To them the collateral ligaments are attached.

33 Feel the **fibular collateral ligament** (lateral ligament) as a cord in front of the biceps tendon from the epicondyle to the head of the fibula (fig. 4.8). The knee must be bent and the leg rotated medially.

34 Map the **tibial collateral ligament** (medial ligament) to the medial surface of the tibia adjacent to the medial border. It is not palpable. Which 3 tendons cross it?

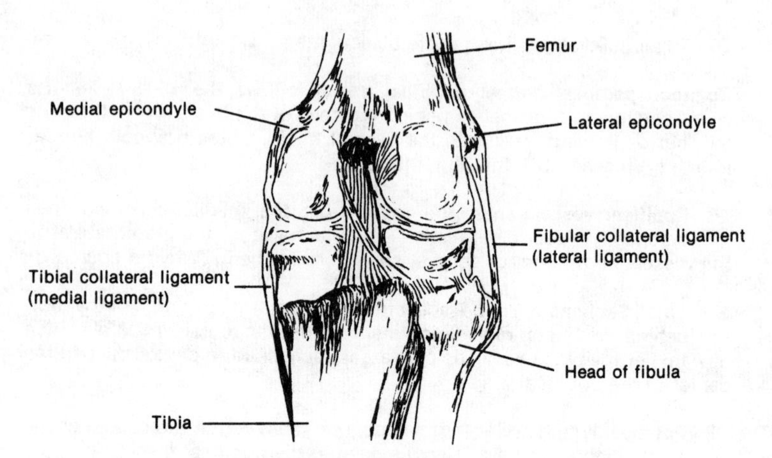

Figure 4.8 (Items 33-34)

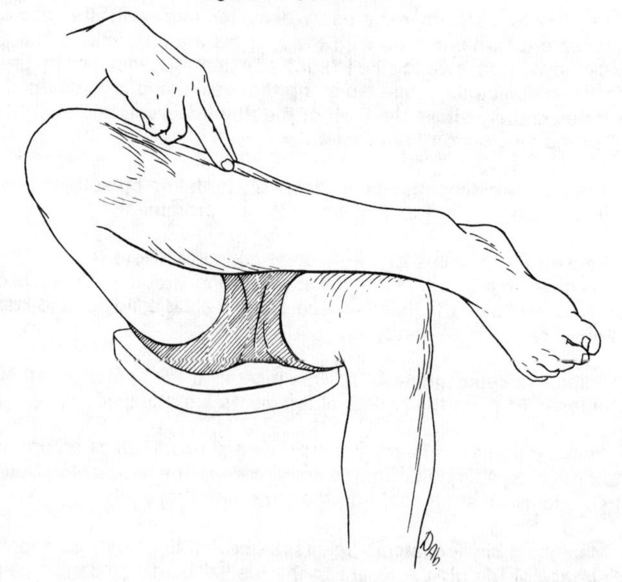

Figure 4.9 (Item 36)

35 **Synovial membrane** extends as a suprapatellar pouch deep to the tendon of quadriceps femoris 3 fingers' breadths above the patella (fig. 4.7). The collateral ligaments lie outside the membrane, hence it must pass below the epicondyles and gastrocnemius and therefore it covers only the anterior third of the sides of the condyles. It does not descend for more than a few mm on the tibia, except where its prolongation deep to the popliteus tendon carries it behind the superior tibiofibular joint. Plot the outline of the synovial membrane on the skin surface.

THE LEG

36 With the ankle resting on the opposite knee (fig. 4.9), palpate the medial surface of the **shaft of the tibia.** Note its continuity with the medial malleolus below and the medial condyle above. Palpate the anterior and medial borders of the shaft and the circumference of the malleolus.

37 Palpate the **head of the fibula.** It lies postero-laterally (fig. 4.10).

38 In contact with the neck of the fibula laterally, is the **common peroneal [lateral popliteal] nerve.** (See No. 25). In contact with the neck medially is the **anterior tibial artery** (not palpable).

39 Run the fingers around the circumference of the **lateral [fibular] malleolus** (fig. 4.11). Palpate its lateral surface which is continuous with the triangular subcutaneous lower end of the fibula. The apex of this triangular lower end is continuous with the anterior peroneal septum.

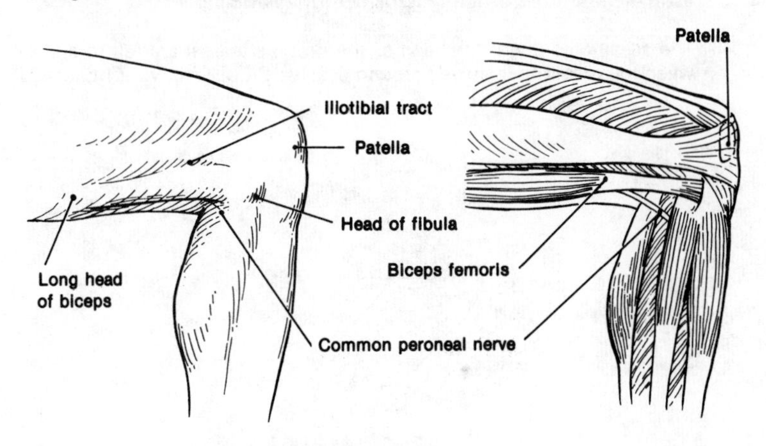

Figure 4.10 (Item 38)

40 Stand on the toes and so bring into prominence the two heads of **gastrocnemius** with **soleus** bulging on each side of it (fig. 4.12). Note also the posterior peroneal septum as a furrow which no nerve crosses, except the common peroneal at the neck of the fibula.

THE FOOT

41 Pressing upwards with two fingers of each hand, feel the blunt lower end of the **medial malleolus** and the sharp end of the lateral malleolus (fig. 4.13). The **lateral malleolus** lies just below and farther back than the medial.

42 **Sustentaculum tali** of the calcaneus is more than a finger's breadth below the medial malleolus. To feel it, you must approach it from below and press upwards with two fingers (fig. 4.14).

43 **Tuberosity of the navicular** is palpable well in front of the sustentaculum. You must press upwards (fig. 4.15).

44 **Head of the talus,** often visible, occupies the space between the sustentaculum tali and the tuberosity of the navicular and is partly supported by the **plantar calcaneonavicular [spring] ligament.**

45 **Base of the first metatarsal** is the length of the cuneiform (easily estimated) in front of the tuberosity of the navicular.

46 The sesamoids under the head of the first metatarsal are felt to slide when the great toe is moved around passively (i.e., with your fingers).

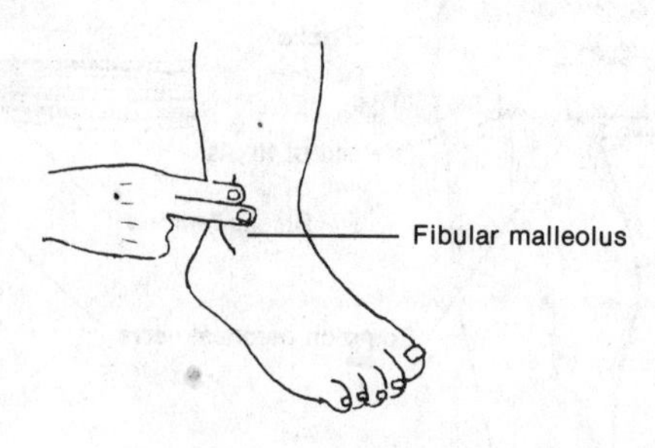

Figure 4.11 (Item 39)

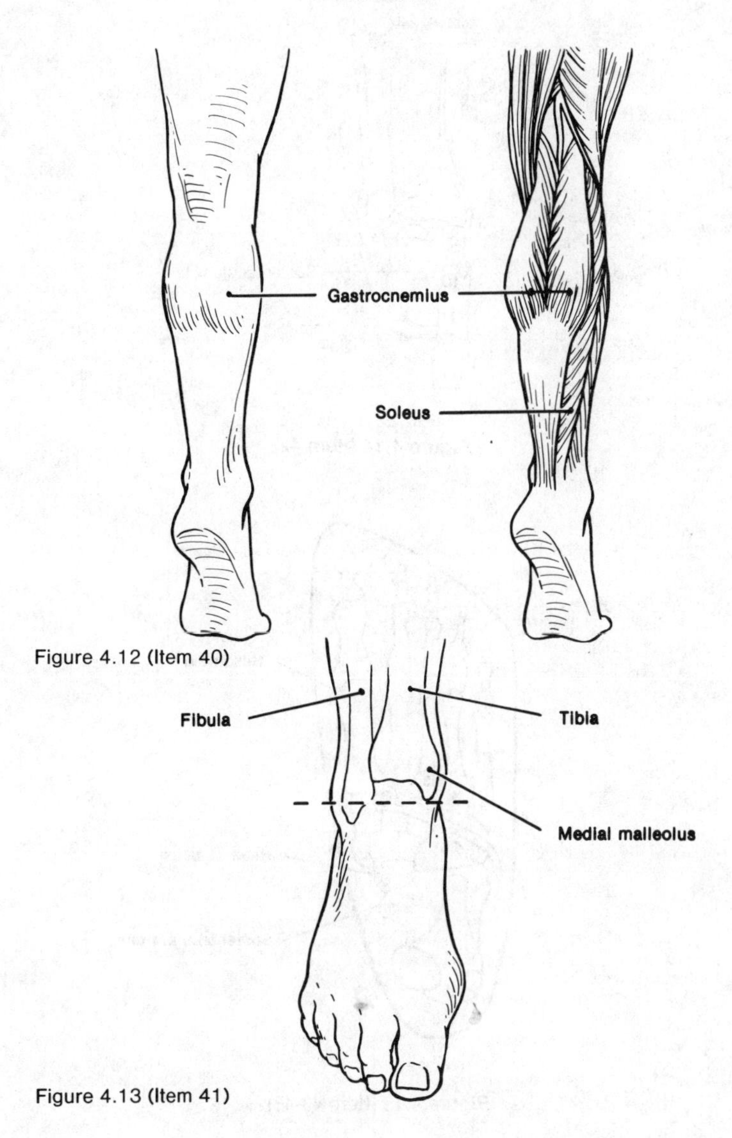

Figure 4.12 (Item 40)

Figure 4.13 (Item 41)

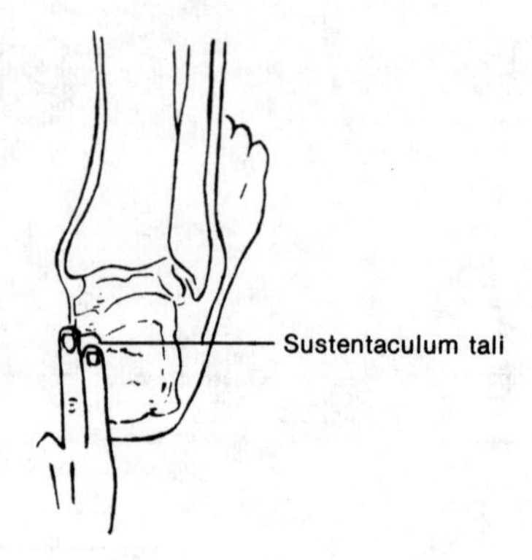

Figure 4.14 (Item 42)

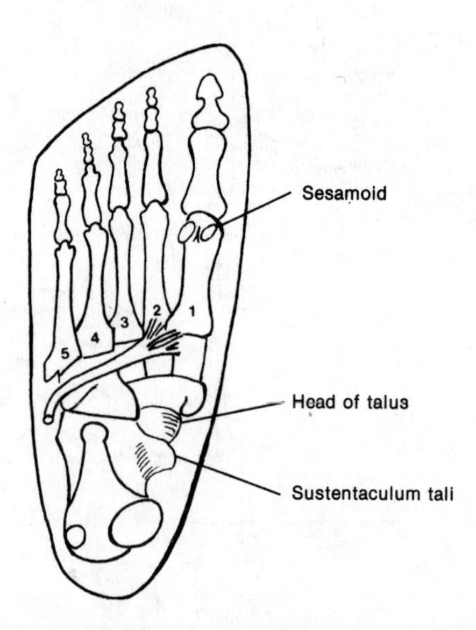

Figure 4.15 (Item 43-48)

47 The tip of the **base of the 5th metatarsal** is easily palpated at the midpoint of the lateral border of the foot on pressing forwards and medially. Its bulge can be seen even through a shoe.

48 **Calcaneocuboid** joint lies ⅔ of the way from the lateral malleolus to the base of the fifth metatarsal. Invert the foot and press backwards with two fingers. You will then readily palpate the uncovered parts of the anterior aspects of the calcaneus and head of the talus, i.e., the **transverse tarsal joint.**

49 **Invertors** (fig. 4.16). With the leg resting on the opposite knee, invert the foot and dorsiflex the ankle, and so bring into prominence **tibialis anterior** and trace it to its insertion on the first metatarsal base and first cuneiform. It is the most anterior structure at the ankle. Invert and plantar-flex and render prominent the tibialis posterior. Trace it to the tuberosity of the navicular (see item 43).

50 Raise the foot from the floor, and dorsiflex and evert it; now trace and palpate the **peroneus brevis tendon** from the lateral malleolus to the base of the 5th metatarsal. Trace and map the **peroneus longus tendon** from the malleolus, below the peroneal tubercle, to the groove on the cuboid behind the 5th metatarsal.

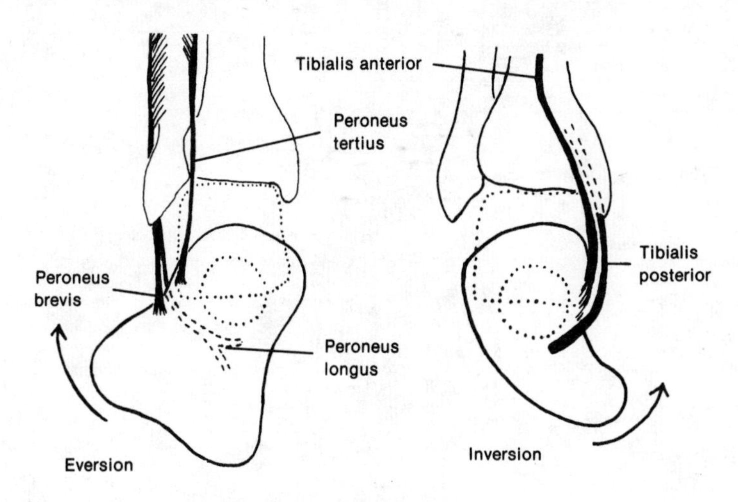

Figure 4.16 (Item 49)

51 **Anterior tibial artery** (fig. 4.17). Map it from the neck of the fibula to midway between the malleoli. The tibialis anterior and extensor hallucis longus muscle lie medially. The extensor digitorum longus and peroneus tertius lie laterally. These tendons are usually conspicuous on dorsiflexion.

52 Trace the **dorsalis pedis artery** from where the anterior tibial artery ends to the proximal end of the first intermetatarsal space. Try to feel its pulse which may be difficult or even impossible because occasionally it is replaced by a large perforating branch of the peroneal artery. Where should you feel for the latter?

53 **Superior extensor retinaculum,** a broad band, joins the anterior border of the tibia to the anterior border of the fibula just above the malleoli. Map it (fig. 4.18).

54 **Extensor digitorum brevis** (fig. 4.19) appears as a soft mass on the dorsum of the foot in front of the lateral malleolus when the foot rests on the floor. Feel it.

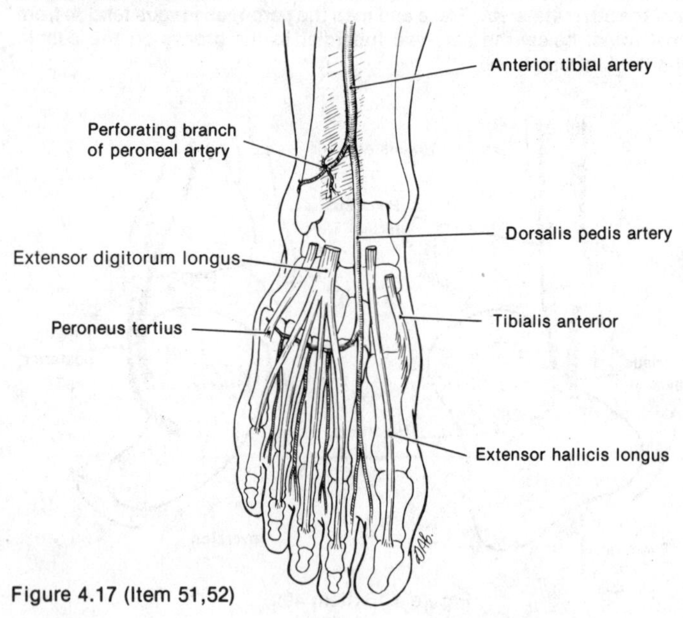

Figure 4.17 (Item 51,52)

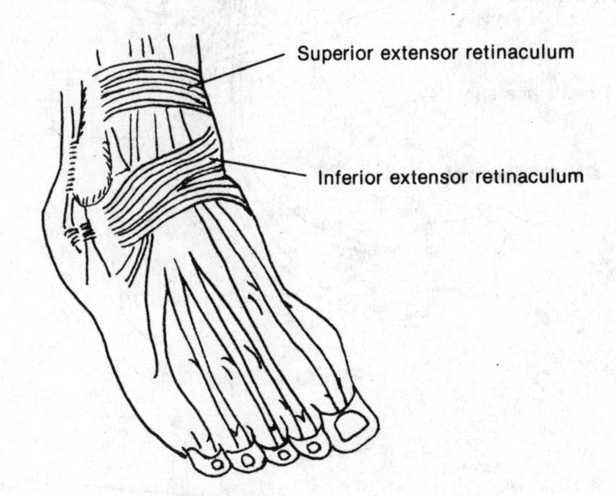

Figure 4.18 (Item 52-53)

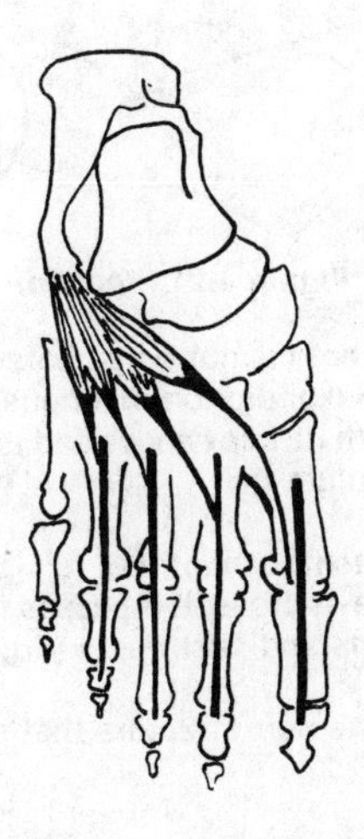

Figure 4.19 (Item 54)

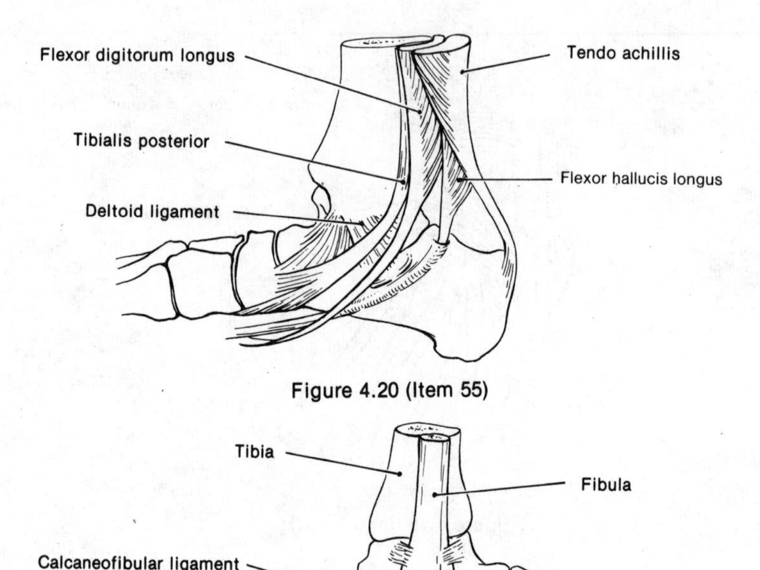

Figure 4.20 (Item 55)

Figure 4.21 (Item 56)

55 **Medial [deltoid] ligament** cannot be palpated because the tendons of tibialis posterior and flexor digitorum longus cross it (fig. 4.20); it may be mapped as a triangle with its apex above and its base extending from the navicular to the sustentaculum tali.

56 **Calcaneofibular ligament** cannot be felt because the tendons of peronei longus and brevis cross it. It passes from just in front of the tip of the malleolus downwards and backwards (fig. 4.21).

57 **Abductor hallucis** is the soft structure that occupies the concavity of the instep (fig. 4.22).

58 **Abductor digiti minimi** is the soft structure that covers (or 'forms') the lateral border of the foot.

59 **Posterior tibial artery** extends from the lower angle of the popliteal fossa to the medial malleolus where it lies the breadth of the tendon of tibialis posterior and flexor digitorum longus behind the medial malleolus; here, with the parts relaxed, its pulsation is usually palpable.

60 Map the usual cutaneous nerve supply of the dorsal and plantar surfaces of the foot. (fig. 4.23)

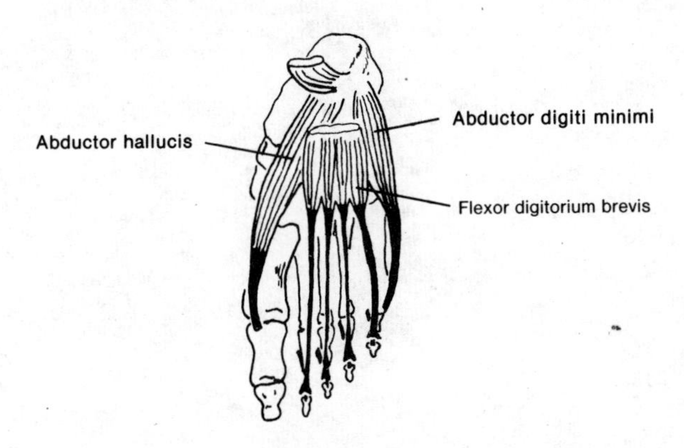

Figure 4.22 (Items 57-58)

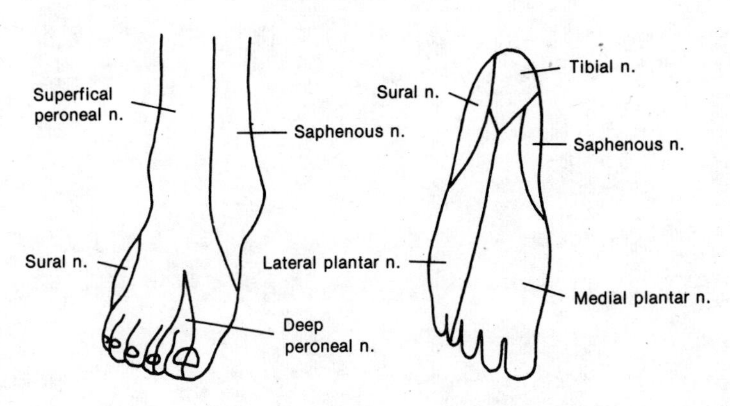

Figure 4.23 (Item 60)

HEAD AND NECK

5

FACE AND CRANIUM FROM THE FRONT

1 **Outline.** The highest visible and palpable part of the cranium is the vertex. From there the outline of the cranium and face seen from the front is called the **norma frontalis.** It ends at the chin (mentum).

2 The frontal region (forehead) and scalp lie above the visible and (easily palpable) superior **margin of the orbit.** The **superciliary arch** above each orbit is much more pronounced in men than in women and in children.

3 Palpate the sharp margins of the orbit and note they are less pronounced medially. At the medial angle of the eye, feel a small subcutaneous lump, the **medial palpebral ligament** (and its bony tubercle), to which the **Orbicularis Oculi** muscle is attached.

4 Four cm from the midline, can you feel a **supraorbital notch** and (in yourself) cause pain by pressing on the emerging supraorbital nerve? Perhaps your notch is closed over to form an impalpable foramen but the pain will still be there, as any anesthetist who uses it to arouse a moribund patient can attest.

5 You really cannot feel the **lacrimal gland** (in the upper lateral corner of the orbital margin) or lacrimal sac (lower medial corner). The eyeball and lids are treated below (items 18-23).

6 **Nose.** Palpate the bridge of the nose (**nasal bones**) and the margins of the anterior nasal aperture (piriform aperture). With the tip of a thumb in a nostril and the index finger outside, confirm that the lower part of the **ala** (wing) is soft tissue and the upper is alar cartilage. With the tip of the thumb in one nostril and index in the other palpate the lower edge of the **cartilaginous septum.**

7 Palpate the **zygoma** (cheek bone). It forms one-third of the orbital margin. Its lower lumpy border gives origin to --

8 **Masseter muscle.** Tense the muscle by clenching the teeth. It covers the **mandibular ramus** down to the **angle of the mandible** (fig. 5.1).

9 In some persons, the **parotid duct** can be rolled under the fingertips as it runs forward from the (inpalpable) **parotid gland** on the surface of the masseter.

10 Follow the **posterior margin** of the mandible with your fingertips up to where it is covered by the cartilage of the external ear.

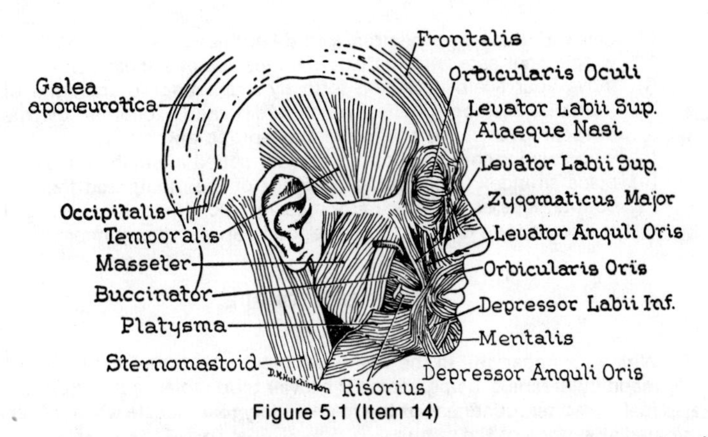

Figure 5.1 (Item 14)

11 Place your little finger in the ear canal (external acoustic meatus) and move your jaw vigorously. The head of the mandible will be felt through the **cartilage of the canal.**

12 Behind the orbital margin, the broad area above the external ear is the **temporal region.** Repeatedly clench your jaw and feel the **temporalis** muscle which fills this region and descends deep to the zygomatic arch to the mandible.

13 **Lips.** Grasp the margin of the upper and then the lower lip between your thumb and index finger. You can feel the pulsations of the **superior** and **inferior labial arteries** (which run between the muscle layer and submucous gland layer of the lips).

FACIAL MUSCLES

14 Looking in a mirror and also while palpating the various regions in which the facial muscles lie, confirm the general locations of muscles of facial expression, e.g.:

(a) the horizontal forehead-wrinkles are caused by **frontalis** (fig. 5.1) elevating the eyebrows.
(b) the 'crowsfeet' at the lateral angles of the eyes are caused by squinting -- **orbicularis oculi.**
(c) the vertical wrinkles between the eyebrows are caused by **corrugator supercilii** and the horizontal wrinkles above the bridge of the nose by **procerus.**

(d) can you dilate your nostrils with **dilator naris? --**
(e) wiggle your ears with the external muscles of the ear? or --
(f) move your scalp back and forth by alternating contractions of **frontalis** and **occipitalis,** which are joined together by the deepest layer of the scalp, the aponeurotic **epicranius?**
(g) wrinkle your chin by contracting your paired **mentalis** muscles.
(h) place an index finger in the **vestibule** of the mouth and feel the contractions of the **buccinator** muscle which lines the cheek and is used during chewing to force food between the grinders.

MOUTH AND TONGUE

15 With your finger still in the vestibule, feel the **anterior margin of the mandibular ramus** rising behind the lower third molar tooth. You may also feel the softer roundness of the **medial pterygoid** muscle which covers the medial surface of the ramus.

16 Behind the upper third molar tooth, feel the **maxillary tubercle** behind and above which lies the **infratemporal fossa,** containing the **lateral pterygoid muscle, mandibular nerve and its branches** and **maxillary** vessels.

17 With a mirror and appropriate use of a tongue depressor or a finger, view in your own mouth (and confirm by inspecting a companion's) --

(a) the vestibule, outside the horseshoe shaped **dental arcades.**
(b) the teeth -- note that the upper incisors overlap the lower ones and that each tooth generally grinds on two teeth of the other arch. Note the cusps.
(c) Can you see saliva emerging from the tiny **orifice of the parotid** duct opposite the upper second molar tooth?
(d) **Oral Cavity Proper.**
  (1) The rough upper surface of the **tongue** is in contrast to the smooth mucosa of the lower surface;
  (2) The blue **sublingual vein** can be seen bilaterally on the inferior surface;
  (3) The **sublingual ridge** on the floor of the mouth on each side of the tongue may be oozing saliva from the multiple ducts of the sublingual salivary gland;
  (4) When the tongue is raised upward, the midline **frenulum** is quite marked; on either side of it a **submandibular salivary duct** empties on the apex of a fleshy papilla.

18 In young subjects the **palatine tonsils** are easily seen on each side of the **fauces** -- the region that divides mouth from pharynx; the ridge runs from the palate to the tongue anterior to the tonsils is the palatoglossal fold.

EYELIDS

19 By looking in a mirror and also inspecting the eye of a companion note that the opposed edges of the eyelids are flattened except for their medial one-fifth which is rounded and hairless.

20 At the medial angle of the eye is a triangular area, the **lacus lacrimalis,** bounded laterally by a free crescentic fold of conjunctiva, the **plica semilunaris.**

21 Gently pull down the lower lid to note a **papilla** at the apex of which is a pin-point **punctum,** the opening of the tiny drainage canal -- the **canaliculus** -- which runs to the lacrimal sac (impalpable).

22 A similar papilla and punctum will be seen in the upper lid.

23 Note the **cilia [eyelashes]** in 2 or 3 irregular rows and perhaps you will see vertical yellowish white streaks through the conjunctiva of the lids -- the **tarsal glands.**

CUTANEOUS SUPPLY OF FACE AND SCALP

24 Fig. 5.2 outlines the areas of the scalp supplied by the **ophthalmic** (V^1), **maxillary** (V^2), and **mandibular** (V^3) branches of the **trigeminal nerve,** and the distribution of **cervical nerves 1 and 2.**

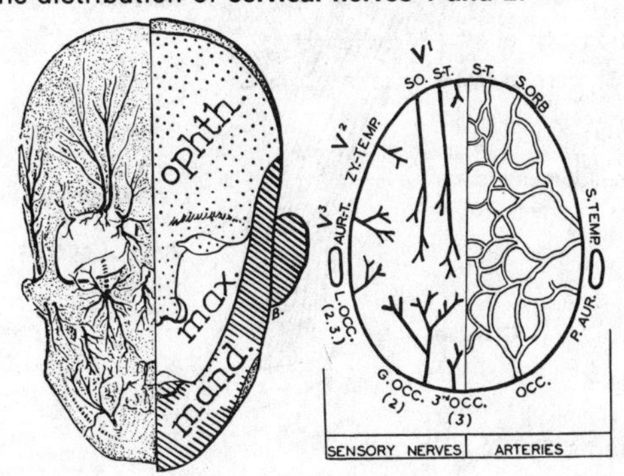

Figure 5.2
(Item 24)

FRONT AND SIDE OF NECK

Midline Structures

25 Elevate the chin and then run the palpating fingers down the midline from chin to suprasternal notch. Feel --

26 The soft **floor of the mouth** formed by **suprahyoid muscles** (and the tongue above them).

27 At the juncture of the floor of the mouth and the neck feel the firm **hyoid bone.** With index and thumb grasp that U-shaped horizontal bone and shuttle it from side to side (fig. 5.3).

28 Swallow. The hyoid bone will move upward pulling upward the --

29 **thyroid cartilage,** the prominence of which is the Adam's apple. Feel the lateral **laminae of the thyroid cartilage.** Below its lower edge feel --

30 the **cricoid cartilage,** firm and rounded in the midline. Below it, flick your fingernail gently down the midline across 3 or 4 **tracheal rings.** Gently grasp the trachea between index finger and thumb and feel it move upward during swallowing. The **thyroid gland** is usually impalpable where its isthmus crosses the trachea in the neck.

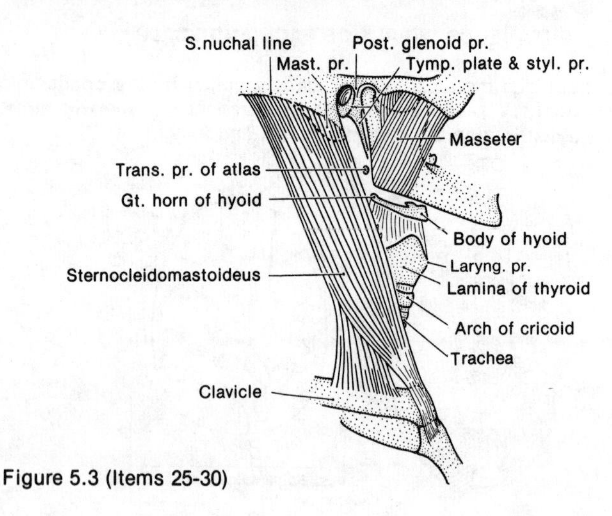

Figure 5.3 (Items 25-30)

More Lateral Structures

31 Feel laterally from the trachea at the suprasternal notch and you will easily identify the **tendon of sternocleidomastoid.** It divides the **anterior from the posterior triangle of the neck,** running upward to the **mastoid process** behind the external ear. To make it most prominent turn the face upward and to the opposite side against resistance.

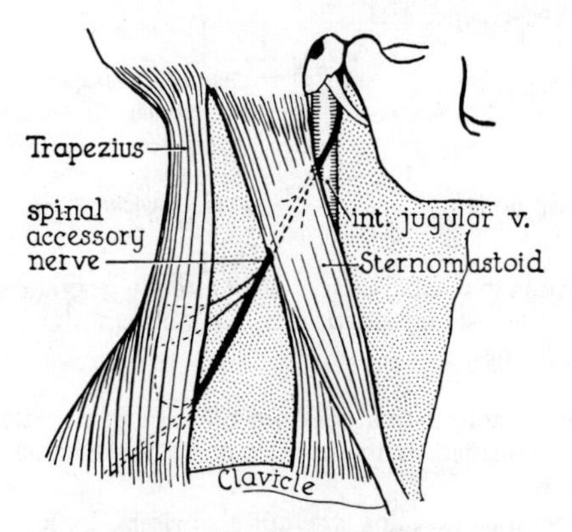

Figure 5.4 (Items 33,34)

32 Between the mastoid process and the angle of the jaw, palpate the **tip of the transverse process of the atlas** (C.1).

Posterior Triangle of Neck

33 The apex of the triangle is about 4 cm behind the ear and its base is the middle one-third of the clavicle. Sternocleidomastoid and trapezius muscles form its oblique borders (fig.5.4).

34 The external branch of the accessory (XI) nerve ('spinal accessory nerve') runs in fascia from deep to sternocleidomastoid near its upper end to deep to trapezius just above the clavicle.

35 **Sensory branches of C2, 3 and 4** radiate to a large area including the ear and side and front of the neck from the point at which the accessory nerve enters the triangle. These nerves are the **lesser occipital, great auricular, anterior cutaneous of the neck** and **supraclavicular nerves.** The last supply skin on the chest wall down to the level of rib 2.

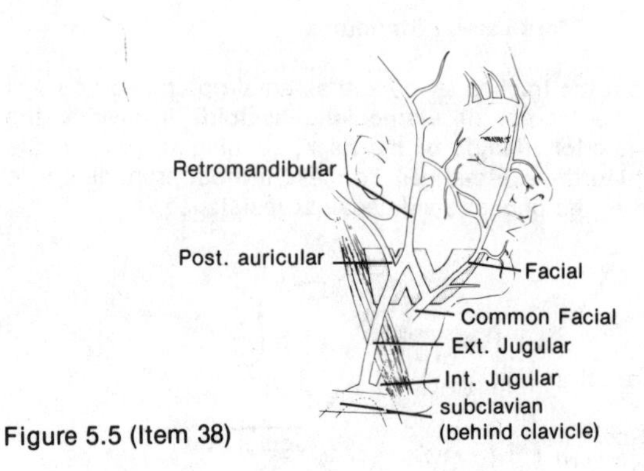

Figure 5.5 (Item 38)

36 **Lymph nodes** that drain deeper structures (e.g. oral cavity, tonsils and pharynx) may be enlarged and palpable in the posterior triangle following infections.

37 **The great vessels in the neck** are difficult to palpate because of the protection afforded by the clavicle and the sternocleidomastoid.

38 The **subcutaneous veins** are quite obvious, especially the **external jugular vein** which descends obliquely across the sternocleidomastoid to plunge deep just above the clavicle (fig. 5.5).

39 The **phrenic nerve** runs vertically downward deep to the lower end of sternocleidomastoid. It lies on the surface of the deep-lying **scalenus anterior** muscle which inserts on the upper surface of the first rib.

40 On the **upper surface of the first rib** behind the insertion of sternocleidomastoid lie the **subclavian vein, subclavian artery** and **brachial plexus** (trunks). Because of the slope of the first rib the order is not only from front to back but also rising upward, the vein being so low as to be 'hidden' entirely by the clavicle. This permits a surgeon to inject the brachial plexus above and behind the vessels on the superior surface of the

41 The **cervicothoracic sympathetic ganglion [stellate ganglion]** lies in front of the neck of the first rib and just above the pleural **cupola.** A needle for injecting it (e.g. with anesthetic) can be introduced above the middle of the clavicle and directed backwards and medially toward the vertebra prominens (fig. 1.3) until the needle strikes the neck of the first rib.

42 **Carotid Arteries.** The **common** carotid is covered completely by sternocleidomastoid. At the level of the upper margin of the thyroid cartilage it divides into **internal** and **external** carotid arteries, the pulsations of which are felt below the angle of the jaw. Easily palpable branches of the external carotid --

(a) **Superficial temporal artery,** on the side of the head (and often visible in elderly thin persons); and
(b) **Facial artery,** crossing the body of the mandible as it enters the face in front of the masseter muscle; it can then be felt by grasping the cheek between the thumb and index about 2 cm behind the angle of the mouth.

43 **External ear:** the main visible and palpable features (fig. 5.6).

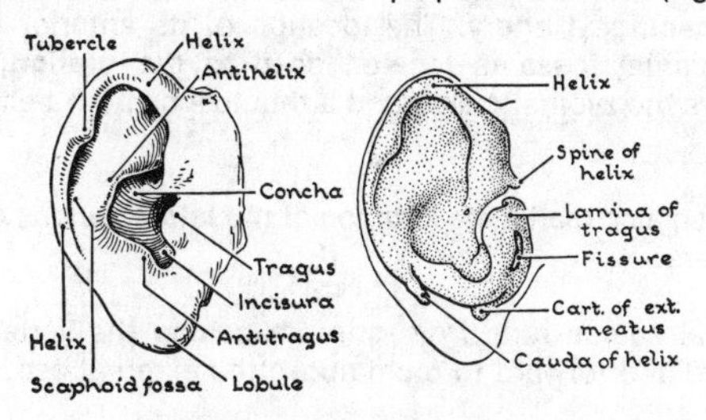

Figure 5.6 (Item 43)

BACK OF HEAD AND NECK

44 From the **vertex**, the **cranial vault** curves downward to the **mastoid process,** bilaterally. Palpate the mastoid processes behing the ears.

45 Run your fingers to the midline posteriorly along the **superior nuchal** lines which separate firm bone (occipital bone) from soft muscles of the **suboccipital region.**

46 At the midline, you will feel the **inion** (external occipital protuberance), an important landmark.

47 Palpate vertically downward in the midline, noting that when the subject flexes his neck (forward bending sharply) the underlying tissue, **ligamentum nuchae,** may raise a firm ridge, but the vertebral spines are indistinctly felt deep to it.

48 Palpating downward in the midline, the first marked prominence you will feel, the **vertebra prominens** [fig. 1.3], is the **seventh cervical spinous process.** (The first thoracic spine is even more prominent in some persons). Have your subject hang his head and these projections are easily seen in profile.

49 The bulk of the bilateral soft muscle mass in the suboccipital region is formed by the **semispinalis capitis** (and **splenius**] covered by the **trapezius.** The superior sloping margin of the trapezius can be palpated down to its insertion on the lateral end of the clavicle. It defines the posterior edge of the posterior triangle of the neck.

INTERIOR OF CRANIUM

50 **Middle meningeal artery.** The location of its anterior branch in the middle cranial fossa is located deep to the **pterion,** two fingers' breadths above the zyomatic arch and a thumb's breadth behind the orbital margin.

51 The pterion also marks the location of the **lateral sulcus** of the cerebral hemisphere.

52 The lateral sulcus runs from one cm behind the vertex of the skull downward and forward to 5 cm above the external acoustic meatus.